Environmental Control for Tuberculosis:

Basic Upper-Room Ultraviolet Germicidal Irradiation Guidelines for Healthcare Settings

DEPARTMENT OF HEALTH AND HUMAN SERVICES
Centers for Disease Control and Prevention
National Institute for Occupational Safety and Health

This document is in the public domain and may be freely copied or reprinted.

DISCLAIMER

Mention of any company or product does not constitute endorsement by the National Institute for Occupational Safety and Health (NIOSH). In addition, citations to Web sites external to NIOSH do not constitute NIOSH endorsement of the sponsoring organizations or their programs or products. Furthermore, NIOSH is not responsible for the content of these Web sites.

ORDERING INFORMATION

To receive documents or other information about occupational safety and health topics, contact NIOSH at

Telephone: **1–800–CDC–INFO** (1–800–232–4636)
TTY: 1–888–232–6348
E-mail: cdcinfo@cdc.gov
or visit the NIOSH Web site at **www.cdc.gov/niosh**.

For a monthly update on news at NIOSH, subscribe to NIOSH eNews by visiting **www.cdc.gov/niosh/eNews**.

DHHS (NIOSH) Publication No. 2009–105

March 2009

SAFER • HEALTHIER • PEOPLE ™

FOREWORD

Tuberculosis (TB) is a severe, contagious disease that primarily affects the lungs. It is usually transmitted from person to person via airborne particles that contain TB bacteria. One of the major concerns for healthcare workers is the potential for transmission of TB at worksites from undiagnosed or unsuspected cases. Upper-room ultraviolet germicidal irradiation (UVGI) systems are considered a supplement or adjunct to other TB infection-control measures (e.g., ventilation) in settings where persons with undiagnosed TB could potentially contaminate the air (e.g., waiting rooms, emergency rooms, corridors, central areas). Also, as an adjunct to negative pressure ventilation, upper-room UVGI can be used in rooms or areas where suspected or confirmed infectious TB patients are isolated or high-risk procedures are performed (e.g., bronchoscopy, sputum induction).

Since 1997, CDC/NIOSH has conducted and funded studies to determine the ability of upper-room UVGI systems to kill or inactivate airborne mycobacteria in a simulated healthcare room. These studies have shown that a properly designed and maintained upper-room UVGI system may be effective in killing or inactivating TB bacteria. Additional research still needs to be done in several areas including fixture installation and room irradiance measurements. However, sufficient laboratory information is now available to develop basic guidelines. This document makes information readily available to occupational safety and health professionals responsible for developing and implementing infection control programs in healthcare settings.

                              Christine M. Branche, Ph.D., Acting Director
                              National Institute for Occupational Safety
                                  and Health
                              Centers for Disease Control and Prevention

ABSTRACT

Although the number of cases of tuberculosis (TB) in the United States has declined in the last several years, there is still a continuing need to protect healthcare workers and the public from risk of infection. One of the primary risks to healthcare workers and the public is exposure to persons with unsuspected or undiagnosed infectious TB. Exposures of this type may occur in areas such as waiting rooms, corridors, or emergency rooms in healthcare facilities (e.g., hospitals, correctional institutions, nursing homes, clinics). While mechanical ventilation systems may provide protection to workers in these situations, there are limitations such as environmental constraints, cost, and comfort considerations. In 1997, the Centers for Disease Control and Prevention (CDC), National Institute for Occupational Safety and Health (NIOSH) awarded a contract to the University of Colorado to evaluate the ability of a well-designed and thoroughly characterized upper-room ultraviolet germicidal irradiation (UVGI) system to kill or inactivate airborne mycobacteria. A number of parameters were evaluated during the study. These included (1) the irradiance level in the upper room that provides a UVGI dose over time that kills or inactivates an airborne surrogate of *Mycobacterium tuberculosis*, (2) how to best measure UVGI fluence levels, (3) the effect of air mixing on UVGI performance, (4) the relationship between mechanical ventilation and UVGI systems, (5) the effects of humidity and photoreactivation (PR), and (6) the optimum placement of UVGI fixtures. The completed research indicates that an appropriately designed and maintained upper-room UVGI system may kill or inactivate airborne TB bacteria and increase the protection afforded to healthcare workers while maintaining a safe level of UVGI in the occupied lower portion of the room. Additional research still needs to be done to better plan effective upper-room UVGI fixture installation considering variables such as air mixing and measurement of the average UV fluence rate in the upper room. However, sufficient laboratory information is now available to develop these guidelines. This document is designed to provide information to healthcare managers, facility designers, engineers, and industrial hygienists on the parameters necessary to install and maintain an effective upper-room UVGI system.

EXECUTIVE SUMMARY

Airborne transmission of *Mycobacterium tuberculosis* is a known risk to healthcare workers. In 1994, the Centers for Disease Control and Prevention (CDC) updated guidelines for preventing the transmission of *M. tuberculosis* in healthcare facilities. The guidelines were issued in response to a resurgence of tuberculosis (TB) that occurred in the United States from the mid-1980s through the early 1990s. The guidelines were based on a risk process that classified healthcare facilities according to categories of risk with a related series of administrative, environmental, and respiratory protection controls.

The TB infection control measures specified in the 1994 CDC document were implemented by many healthcare facilities in the United States. These measures helped decrease the transmission of *M. tuberculosis* to patients and healthcare workers in healthcare facilities. However, despite the decline in TB rates in the United States in recent years, there is still a continuing need to protect healthcare workers from the risk of infection. For example, healthcare workers in different areas face varied risks. In Idaho, the case rate was 0.8 per 100,000 in 2004, while in the District of Columbia the rate was 14.6 per 100,000.

The 1994 CDC guidelines were primarily aimed at hospital-based facilities. In 2005, CDC reevaluated the risk of TB transmission to healthcare workers and developed new guidelines. These guidelines are intended to eliminate the lingering threat to healthcare workers that is primarily from patients or others who have unsuspected and undiagnosed infectious TB. Exposures of this type may occur in areas such as waiting rooms, corridors, or emergency rooms in healthcare facilities (e.g., hospitals, correctional institutions, nursing homes, clinics).

The use of ultraviolet germicidal irradiation (UVGI) in the upper portion of rooms or areas is an environmental control measure that may be effective in killing or inactivating airborne *M. tuberculosis* generated by persons with unknown or unsuspected infectious TB. There was concern regarding the efficacy of this control measure when CDC updated the healthcare facility guidelines in 1994. Therefore, in 1997, the CDC's National Institute for Occupational Safety and Health (NIOSH) awarded a contract to the University of Colorado to evaluate the ability of a well-designed and thoroughly characterized upper-room UVGI system to kill or inactivate an airborne surrogate of *M. tuberculosis*. A number of parameters were evaluated during the study. These included (1) the UV fluence rate needed to inactivate *M. tuberculosis* surrogates, (2) the way to best measure UVGI fluence rates, (3) the effect of air mixing on UVGI performance, (4) the relationship between mechanical ventilation and UVGI systems, (5) the effects of humidity and photoreactivation (PR), and (6) the optimum placement of UVGI fixtures. The completed research and other research studies that have recently been published clearly indicate that an appropriately designed and maintained upper-room UVGI system may kill or inactivate airborne TB bacteria and significantly increase the protection afforded to healthcare workers while maintaining a safe level of UVGI in the occupied lower portion of the room. Additional research needs to be done to better plan effective upper-room UVGI fixture installations in healthcare settings considering variables such as air mixing and measurement of UV fluence levels in the upper room. However, sufficient laboratory information is now available to develop basic guidelines.

The purpose of this document is to examine the different parameters necessary for an effective upper-room UVGI system and to provide this information to occupational safety and health professionals responsible for developing and implementing infection control programs in healthcare settings. These guidelines are consistent with previous CDC healthcare guidelines and expand upon them.

Guidelines

Factors Influencing Effectiveness of Upper-Room UVGI Systems

1. UVGI Irradiance and Dose

Factors that must be considered when evaluating the ability of an upper-room UVGI system to kill or inactivate airborne microorganisms include the sensitivity of the microorganisms to UVGI and the dose of UVGI received by a microorganism or population of microorganisms. UVGI dose is the ultraviolet (UV) irradiance multiplied by the time of exposure and is usually expressed as $\mu W \cdot s/cm^2$.

A well-designed upper-room UVGI system may be effective in killing or inactivating most airborne droplet nuclei containing mycobacteria if designed to provide an average UV fluence rate in the upper room in the range of 30 $\mu W/cm^2$ to 50 $\mu W/cm^2$, provided the other elements stipulated in these guidelines are met. In addition, the fixtures should be installed to provide as uniform a UVGI distribution in the upper room as possible.

2. Upper-Room UVGI Systems and Mechanical Ventilation

As the mechanical ventilation rate in a room is increased, the total number of microorganisms removed from the room via this system is increased. However, when mechanical ventilation is increased in a room where an upper-room UVGI system has been deployed, the effectiveness of the UVGI system may be reduced because the residence time of the bacteria in the irradiated zone decreases.

Under experimental laboratory conditions with mechanical ventilation rates up to six air changes per hour (ACH), the rate that microorganisms are killed or inactivated by UVGI systems appears to be additive with mechanical ventilation systems in well-mixed rooms.

3. Air Mixing

Upper-room UVGI systems rely on air movement between the lower portion of the room where droplet nuclei are generated and the upper irradiated portion of the room. Once in the upper portion, droplet nuclei containing *M. tuberculosis* may be exposed to a sufficient dose of UVGI to kill or inactivate them.

When upper-room UVGI systems are installed, general ventilation systems should be designed to provide optimal airflow patterns within rooms and prevent air stagnation or short-circuiting of air from the supply diffusers to the exhaust grills. Also, heating and cooling seasons should be considered and the system designed to provide for optimal convective air movement.

Most rooms or areas with properly installed supply diffusers and exhaust grills should have adequate mixing. If areas of air stagnation are present, air mixing should be improved by adding a fan or repositioning the supply diffusers and/or exhaust grills. If there is any question about vertical air mixing between the lower and upper portions of the room due to environmental or other factors, a fan(s) should be used to continually mix the air. In a room without adequate air mixing under experimental laboratory conditions, the UVGI system effectiveness increased from 12% to 89% when a mixing fan was used.

4. Humidity

A number of studies have indicated that the effectiveness of upper-room UVGI systems decreases as humidity increases. The reason for the decrease in UVGI effectiveness is not clearly understood. However, the effect needs to be considered in the general context of upper-room UVGI systems.

For optimal efficiency, relative humidity (RH) should be controlled to 60% or less if upper-room UVGI systems are installed. This is consistent with American Institute of Architects (AIA) and the American Society of Heating, Refrigerating, and Air-Conditioning Engineers (ASHRAE) recommendations that the RH affecting patient care areas in hospitals and outpatient facilities range from 30% RH to 60% RH. If high humidity conditions are normal, it may be necessary to install a system with greater than normal upper-room irradiance levels.

5. Temperature

Recommendations developed by ASHRAE and AIA stipulate that the design temperature for most areas affecting patient care in hospitals and outpatient facilities range from 68 °F to 75 °F (20 °C to 24 °C). This temperature range is consistent with the optimal use of low-pressure mercury lamps that are used in upper-room UVGI systems.

Practical Guidelines for Installation of Upper-Room UVGI Systems

1. UV Lamps

The most common way to generate germicidal UV radiation in lamps used in well-designed upper-room UVGI systems is to pass an electrical charge through low-pressure mercury vapor that has been enclosed in selected glass tubes that transmit only certain

UV wavelengths. Care must be used in selecting the correct UVGI lamp for use in upper-room UVGI systems. Typically, the optimal wavelength for UV germicidal radiation is 254 nanometers (nm) in the UV-C range. UV lamps are made for a variety of purposes that may have a negligible consequence in killing airborne microorganisms. Some UV lamps (such as those used for tanning) radiate energy in the UV-A and/or UV-B range and over extended periods may have adverse health consequences for exposed persons. Other UV lamps are designed to emit radiation at 184.9 nm and produce ozone, which is hazardous to humans even at low concentrations. Low-pressure mercury lamps should be rated for low or no ozone generation. Since all lamps must eventually be discarded, each lamp should contain only a relatively small quantity of mercury (i.e., 5 mg or less).

2. UVGI Fixtures

In upper-room UV irradiation, fixtures containing UVGI lamps are suspended from the ceiling or installed on walls. The base of the lamp is shielded to direct the radiation upward and outward to create an intense zone of UVGI in the upper air while minimizing the level of UVGI in the lower (occupied) portion of the room or area. The height of the room must be considered to design an effective system. Only well-designed fixtures as noted in this document should be used.

3. System Installation

Several rules of thumb for installation of the fixtures for upper-room UVGI systems have been developed over the last 50 to 60 years. In the CDC/NIOSH-funded study as indicated above, a well-designed upper-room UVGI system may be effective in killing or inactivating airborne *M. tuberculosis* if designed to provide an average UV fluence rate in the upper irradiated zone in the range of 30 $\mu W/cm^2$ to 50 $\mu W/cm^2$ provided the other elements stipulated in these guidelines are met. Based on this, two additional rules of thumb (guidelines) are provided in the document for installing UVGI systems. To simplify the installation process, the new guidelines are based on the required UV lamp wattage for the system. Considering all parameters, the installation of UVGI fixtures in rooms with approximately 2.4 m (8 ft) ceilings that provide (1) a UV-C irradiance of 1.87 W/m^2 (0.17 W/ft^2) or (2) a UV-C power distribution of 6 W/m^3 (0.18 W/ft^3) in the upper UVGI zone should be effective in killing or inactivating airborne mycobacteria.

A professional who is knowledgeable in upper-room UVGI systems and system installation should be consulted before procurement and installation of the system. The number of persons properly trained in the design of upper-room UVGI systems is currently limited. Persons who may be consulted include engineers, industrial hygienists, and radiation/health physicists. A mechanism to provide training certification for system designers should be developed.

4. Installation and Maintenance Considerations

Once the number and types of UVGI fixtures appropriate for the room or area have been determined, the fixtures need to be appropriately installed. Installation guidelines are provided in the document as well as problem areas that have been encountered during CDC/NIOSH evaluations. Only qualified service technicians who have received training on the installation and placement of UVGI lamp fixtures should install the systems.

Discussions are provided in the document on the required maintenance for the UVGI fixtures, UV lamps, and personal protective equipment (PPE) necessary during maintenance. Methods for UVGI measurements in the lower (occupied) level of a room or area and the upper irradiated area are discussed.

Additional research needs for determining the most effective upper-room UVGI systems are provided in the document. These include UVGI measurements, air mixing, the effect of low humidity, microbial sensitivity, and testing and validating upper-room UVGI systems. Also, research needs to be done on the ability of UVGI systems to kill or inactivate microorganisms in (1) different size respirable droplet nuclei and (2) droplet nuclei coated with actual or simulated sputum.

Contents

DISCLAIMER ii
ORDERING INFORMATION ii
FOREWORD iii
ABSTRACT iv
EXECUTIVE SUMMARY v
ABBREVIATIONS xiii
GLOSSARY xv
ACKNOWLEDGMENTS xxi

1 INTRODUCTION 1

2 BACKGROUND 3
    2.1 UVGI Systems 4
    2.2 UV Radiation 4
    2.3 Cellular Effects 5
    2.4 Health Issues 5
    2.5 Exposure Guidelines 5

3 FACTORS INFLUENCING UVGI EFFECTIVENESS 9
    3.1 UV Irradiance and Dose 9
        3.1.1 Sensitivity of Microbial Populations to UVGI 9
        3.1.2 Irradiance—Experimental Results 13
        3.1.3 Basis for the UVGI Irradiance and Lamp Configuration
             Guidelines 16
    3.2 Upper-Room UVGI Systems and Mechanical Ventilation 17
    3.3 Air Mixing 20
        3.3.1 Temperature and Mechanical Ventilation Considerations 20
        3.3.2 Placement of Supply Diffusers and Exhaust Grills 21
        3.3.3 Evaluating Air Mixing 22
    3.4 Humidity 23
        3.4.1 Photoreactivation and Relative Humidity 24
        3.4.2 Basis for Humidity Guidelines 25
    3.5 Temperature 26

4 PRACTICAL GUIDELINES FOR INSTALLING UPPER-ROOM UVGI SYSTEMS 29

   4.1 UV Lamps 29
       4.1.1 Other Considerations 30
       4.1.3 Disposal 31
   4.2 UVGI Fixtures 31
   4.3 System Installation 32
   4.4 Installation and Lamp Maintenance 34
   4.5 UV Fixtures and Lamp Maintenance 36
   4.6 Personal Protective Equipment (PPE) 37
   4.7 UVGI Measurements 38
       4.7.1 Measurement Approaches 39
   4.8 Measuring UVGI Irradiance 39
       4.8.1 Lower (Occupied Level) 39
       4.8.2 Upper-Room (Irradiated) Level 40

5 RESEARCH NEEDS 43

   5.1 UVGI Measurements 43
       5.1.1 Upper-Room Measurements 43
       5.1.2 Lower-Room Measurements 43
   5.2 Room Air Mixing 43
   5.3 Relative Humidity 43
   5.4 Microbial Sensitivity 44
   5.5 Safety and Health Guidelines 44
   5.6 Testing and Commissioning 44
   5.7 Performance Guidelines 44
   5.8 Mechanical Ventilation 44
   5.9 Planning Guidelines 44
   5.10 Photoreactivation 45
   5.11 Temperature 45
   5.12 UVGI Effectiveness 45
   5.13 UV Fixtures 45

REFERENCES 47

APPENDIX A — FIELD PROBLEMS NOTED IN SOME UVGI SYSTEMS 59

APPENDIX B — RESEARCH TOOLS 60

ABBREVIATIONS

ACGIH	American Conference of Governmental Industrial Hygienists
ACH	air changes per hour
AIA	American Institute of Architects
AIDS	acquired immune deficiency syndrome
ASHRAE	American Society of Heating, Refrigerating, and Air-Conditioning Engineers
BCG	Bacillus Calmette-Guérin [*Mycobacterium bovis* BCG]
B. subtilis	*Bacillus subtilis*
°C	degree(s) Celsius
CDC	Centers for Disease Control and Prevention
CFD	computational fluid dynamics
CG	constant generation
cm	centimeter(s)
CMD	count median diameter
DNA	deoxyribonucleic acid
eACH	equivalent air changes per hour
e.g.	for example
E. coli	*Escherichia coli*
°F	degree(s) Fahrenheit
ft	foot/feet
g/cm^3	gram per cubic centimeter
h	hour(s)
He	helium
HIV	human immunodeficiency virus
i.e.	that is
IESNA	Illuminating Engineering Society of North America
in.	inch(es)
KI	potassium iodide
m	meter(s)
mg	milligram(s)
min	minute(s)
mJ/cm^2	millijoule per square centimeter
MSDS	material safety data sheet
M. africanum	*Mycobacterium africanum*
M. bovis BCG	*Mycobacterium bovis* BCG
M. marinum	*Mycobacterium marinum*
M. parafortuitum	*Mycobacterium parafortuitum*

M. phlei	*Mycobacterium phlei*
M. tuberculosis	*Mycobacterium tuberculosis*
NIOSH	National Institute for Occupational Safety and Health
nm	nanometer(s)
NTIS	National Technical Information Service
OSHA	Occupational Safety and Health Administration
PET	permissible exposure time(s)
PPE	personal protective equipment
PR	photoreactivation
REL	recommended exposure limit
RH	relative humidity
s	second(s)
SF_6	sulfur hexafluoride
S. marcescens	*Serratia marcescens*
S. mitis	*Streptococcus mitis*
TB	tuberculosis
UV	ultraviolet
UV-A	long wavelengths, range: 315 nm to 400 nm
UV-B	midrange wavelengths, range: 280 nm to 315 nm
UV-C	short wavelengths, range: 100 nm to 280 nm
UVGI	ultraviolet germicidal irradiation
W	watt(s)
WHO	World Health Organization
W/m^2	watts per square meter
$\mu J/cm^2$	microjoules per square centimeter
μm	micrometer(s)
$\mu W/cm^2$	microwatts per square centimeter
$\mu W \cdot s/cm^2$	microwatt-seconds per square centimeter

GLOSSARY

Aerodynamic diameter: The diameter of a sphere with a density of 1 g/cm^3 and with the same velocity (due to gravity) as the particle of interest. The aerodynamic behavior of a particle is one of the main factors that determines particle deposition into the human respiratory system. Normally, particles such as droplet nuclei are considered potentially respirable if they have an aerodynamic diameter less than 5 μm.

Air changes per hour (ACH): The total volume of air that flows through a room in 1 h (cubic meters per hour) divided by the room volume in cubic meters.

$$A = Q/V,$$

where A is air changes per hour,
Q is the air volume that flows through the room in 1 h, and
V is the volume of the room.

Air mixing: The degree to which air entering a room moves between areas within the room may be expressed as a mixing factor (K) [ACGIH 2004]. If $K = 1$, the air is well mixed (instantaneous and complete) and the air in the room is spatially uniform. When $K > 1$, nonuniform mixing and spatial variations exist. Values of K could theoretically range from perfect mixing of 1 to poor mixing of 10. However, it is most common to have K-values from 1 to 3 in most office-type settings.

Average lamp life: The average ultraviolet germicidal irradiation (UVGI) lamp life refers to the electrical lifetime of the lamp and is based on the point in time when one half of a statistically large number of lamps fail to light. Lamp manufacturers base their average UVGI lamp lifetimes by cycling the lamps off/on. For example, Philips bases their average lamp lifetime at 1 switch per 3 hours [Philips 2006]. In general, the average lamp life is longer than the average effective lamp life (see also average effective lamp life).

Average effective lamp life: The effective (useful) life of a UVGI lamp may be calculated by the manufacturer from a statistically large number of lamps between the first use (usually following a 100 h break-in period) to a point where the ultraviolet (UV) output in one half of the lamps has declined to a specified level. For example, Philips [2006] considers the effective lamp life to be 9,000 h with 20% decline in the UV output (see also average lamp life).

Bacillus Calmette-Guérin (BCG) [Mycobacterium bovis BCG]: A relatively safe bacterium that is commonly used as a surrogate for virulent *Mycobacterium tuberculosis* in experimental studies. BCG is used in some countries as a vaccine for tuberculosis.

Chemical actinometry: The technique of measuring radiation dose by the change produced in a chemical.

Count median diameter: The diameter in a population (e.g., microorganisms) above which there are as many particles with larger diameters as there are particles with smaller diameters [IAEA 1978].

Decay rate constant: See UV rate constant.

Dose: A term used to describe the amount of radiant energy absorbed in a unit volume, organ, or person. The dose of UVGI received by a microorganism (e.g., in droplet nuclei) is generally assumed to be equivalent to the fluence or UVGI irradiance multiplied by the time of exposure and is usually expressed as $\mu W \cdot s/cm^2$ or $\mu J/cm^2$.

Dose response: The correlation between the amount of energy received by a population of microorganisms and the resulting effect (e.g., inactivation).

Droplet nuclei: Airborne particles that contain bacteria or viruses capable of transmitting disease from person to person [Benenson 1990]. These particles are usually formed from the evaporation of fluids that are emitted from an infected host when they cough, sneeze, or talk. Most droplet nuclei are believed to have an aerodynamic diameter (see definition above) less than 5 μm [Duguid 1946; Fennelly et al. 2004; Papineni and Rosenthal 1997; Wells 1955]. Due to their small size, droplet nuclei may remain airborne for long periods [Loudon et al. 1969] and be inhaled deep into the alveoli of the lungs [Riley and O'Grady 1961].

Equivalent air changes per hour (eACH): The ability of an environmental control (e.g., upper-room UVGI) to kill or inactivate an airborne microorganism at the same rate as mechanical ventilation removes the airborne microorganism from a room as measured in ACH. This is a measure of the UVGI efficacy that can be obtained using decay model (see *Experimental models*) experimental conditions in a well-mixed room. The eACH for different microorganisms may vary.

Experimental models: Wells and Wells [1936] described two general models for evaluating infectious microorganisms generated in a room. They referred to these two models as (1) die-away (i.e., decay model) and (2) equilibrium (i.e., constant-generation model). These models were further described by Miller and Macher [2000] as the following:

Decay model: In a typical decay model, an infectious person has been present in a room for some time and departs. Since the infectious person has left the room, the infectious aerosol concentration decreases with time. Experiments that use this model require that the measurement of microbial concentrations be taken after the generation of bioaerosols has stopped and the room is well mixed [Miller et al. 2002].

Constant-generation (CG) model: In this model, an infectious person has been present in the room for an extended period and remains in the room. The release of infectious particles is essentially constant since the number within the room will increase until the number removed equals the number of particles added in a unit of time assuming a uniform generation rate from the infectious person. In the experimental protocol using this model, bioaerosols are constantly generated during microbial measurements. The room does not necessarily have to be well mixed to interpret the data as it does in the decay model.

Fluence: (also called UV dose) which is similar in construction to "UV rate constant (k) (decay rate constant) below. The radiant energy in µW·s/cm^2 or µJ/cm^2 passing through a surface area. The term fluence differs slightly from UV dose since use of the latter implies total absorption of UV energy whereas fluence represents irradiation only. In this document, the fluence rate is defined as the average UV fluence rate impacting particles in the irradiated zone of the upper-room measured in µW/cm^2.

Flux density: See Radiant flux.

Germicidal: Capable of killing or inactivating microorganisms.

Infection: A condition in which a biological agent capable of causing disease in a susceptible host enters the body and elicits a response from the host's immune system. Infection with *M. tuberculosis* may or may not lead to active disease.

Infectious: Capable of transmitting disease.

Intensity: A general term often used to describe the irradiance of a light source. The term irradiance is used in this document to describe the UV irradiation in units of µW/cm^2 or W/m^2.

Inverse square law: This law suggests that the amount of radiation passing through an area is inversely proportional to the square of the distance of that area from the energy source.

Mathematically, the *Inverse Square Law* is described by the equation,

$$I_2/I_1 = (d_1/d_2)^2,$$

where I_1 is the irradiance at distance d_1 from the energy source and I_2 is the irradiance at distance d_2 from the energy source.

Irradiance: The density of radiation incident on a flat surface. A measure of radiometric flux per unit area that is typically expressed in µW/cm^2.

Mechanical or forced ventilation: Ventilation through an air handling unit or direct injection to a space by a fan that is used to control indoor air quality. A local exhaust fan can enhance infiltration or natural ventilation, thus increasing the ventilation air flow rate.

Mycobacterium parafortuitum: A relatively safe mycobacteria species that has similar biological characteristics found in the pathogenic mycobacteria species, *M. tuberculosis*.

Mycobacterium tuberculosis: A bacterium that causes clinically active symptomatic tuberculosis (TB).

Natural ventilation: Occurs when the air in a space is changed with outdoor air without the use mechanical systems, such as a fan. Most often natural ventilation is assured though operable windows but it can also be achieved through temperature and pressure differences between spaces.

Pathogenic microorganisms: Microorganisms capable of producing disease.

Photoreactivation (PR): Microorganisms (like most living organisms) have an array of mechanisms to repair the damage to their deoxyribonucleic acid (DNA) caused by exposure to UV radiation. In bacteria, these DNA repair methods are often referred to as light and dark mechanisms. The light repair mechanisms (also known as PR) require exposure of the cell to visible light (380 nm to 430 nm) following damage caused by UVGI exposure and the presence of a photoreactivating enzyme known as photolyase. The dark repair mechanisms do not require the presence of visible light as a cofactor [Miller et al. 1999].

Radiometer: An instrument that measures radiation over a wavelength range.

Radiant flux: A measure of radiometric power. The flux of a quantity is a measure of the rate of the quantity flow per unit time passing through a unit area (e.g., W/m^2).

Recommended exposure limit (REL): The Centers for Disease Control and Prevention (CDC), National Institute for Occupational Safety and Health (NIOSH) develops RELs for preventing disease and hazardous conditions in the workplace. The recommendations are then transmitted as formal publications to the Occupational Safety and Health Administration (OSHA) or the Mine Safety and Health Administration (MSHA) of the U.S. Department of Labor (DOL) for use in promulgating legal standards. In 1972, CDC/NIOSH published an REL for UV radiation to prevent adverse effects on the eyes and skin [NIOSH 1972] (see also section 2.5).

Relative humidity (RH): The ratio of the amount of water vapor in the atmosphere to the amount necessary for saturation at the same temperature. RH is expressed in terms of percent and measures the percentage of saturation. At 100% RH, the air is saturated. The RH decreases when the temperature increases without changing the amount of moisture in the air.

Spectrophotometer: An instrument that measures the absorption of photons in a sample. A spectrophotometer may use spectral filters to measure the UV range of photon absorption.

Tuberculosis (TB): A condition caused by infection with a member of the *M. tuberculosis* complex that has progressed to causing clinical illness (manifesting symptoms or signs) or subclinical illness (early stage of disease in which signs or symptoms are not present, but other indications of disease activity are present). The bacteria can attack any part of the body, but disease is most commonly found in the lungs (pulmonary TB). Pulmonary TB disease can be infectious, whereas extrapulmonary disease (occurring at a body site outside the lungs) is not infectious, except in rare circumstances. When the only clinical finding is specific chest radiographic abnormalities, the condition is termed "inactive TB" and can be differentiated from active TB disease, which is accompanied by symptoms or other indications of disease activity (e.g., the ability to culture reproducing TB organisms from respiratory secretions or specific chest radiographic finding) [CDC 2005a].

Ultraviolet germicidal irradiation (UVGI): The use of ultraviolet radiation to kill or inactivate microorganisms.

Ultraviolet germicidal irradiation (UVGI) effectiveness (Efficiency): A measure of the ability of an upper-room UVGI system to kill or inactivate microorganisms. This may be expressed as either eACH in decay experiments or the percentage of microorganisms killed or inactivated by UVGI in CG experiments. This latter measure of effectiveness may be expressed by the following equation:

$$E_{UV} = 100 \times (1 - C_{UV}/C_0),$$

where E_{UV} represents the effectiveness of UVGI as a percentage,
C_{UV} is the concentration of culturable microorganisms with UVGI exposure, and
C_0 is the concentration of culturable microorganisms without UVGI exposure.

Ultraviolet germicidal irradiation (UVGI) lamp: Low-pressure mercury lamps, medium-pressure mercury lamps, and pulsed-UV lamps have been shown to have germicidal activity (see section 4.1). However, only low-pressure mercury lamps have been evaluated in upper-room UVGI systems. Therefore, in the context of this document, a UVGI lamp refers to a low-pressure mercury lamp that kills or inactivates microorganisms by emitting UV germicidal radiation, predominantly at a wavelength of 254 nm. UVGI lamps can be used in ceiling, wall, or corner fixtures in the upper portion of rooms or areas to kill or inactivate airborne microorganisms such as *M. tuberculosis* in droplet nuclei. UVGI lamps can also be used in air ducts of ventilation systems and in portable or fixed air cleaners.

UV rate constant (k) (decay rate constant): The slope of a survival curve for a microbial species exposed to UVGI. It has been suggested that k is directly related to the sensitivity of a microbial species to UVGI and is unique to each species [Kethley 1973]. The UV rate constant is also called the decay rate constant and is virtually equivalent to the Z-value.

Ventilation: The process of changing or replacing air in any space to remove moisture, odors, smoke, heat, dust, and airborne contaminants. Ventilation includes both the exchange of air to the outside as well as circulation of air within the building. Methods for ventilating a building may be divided into mechanical/forced and natural types.

Virulence: The degree of pathogenicity of a microorganism as indicated by the severity of the disease produced and its ability to invade the tissues of a host. *M. tuberculosis* is a virulent organism [CDC 1994].

Z-value: The Z-value [Kethley 1973] represents the ratio of the inactivation rate normalized by UV irradiance. Theoretically, the higher the Z-value for a target microorganism, the greater the susceptibility to UVGI and the more quickly the microorganism will be killed or inactivated. The Z-value response is described by the following equation:

$$Z = \ln(N_0/N_{UV})/D,$$

where N_0 is the number of surviving microorganisms with no UVGI exposure,
N_{UV} is the number of surviving microorganisms following UV exposure, and
D is the UVGI dose in $\mu W \cdot s/cm^2$.

ACKNOWLEDGMENTS

This document was prepared under the auspices of the Division of Applied Research and Technology. John Whalen was the principal author of the document. Larry Reed and Scott Earnest were instrumental in providing direction and ensuring completion of this document. A Task Force provided the wide range of expertise necessary to develop the document. The Task Force members were:

G. Scott Earnest, Ph.D., PE, CSP, Division of Applied Research and Technology
Leroy Mickelsen, M.S., PE, U.S. Environmental Protection Agency (Research Triangle Park, NC)
Gene Moss, Ph.D., Corning, Inc. (Corning, NY)
Laurence Reed, M.S., Division of Surveillance, Hazard Evaluations, and Field Studies
Teresa Seitz, M.P.H., CIH, Division of Surveillance, Hazard Evaluations, and Field Studies
Jennifer Topmiller, M.S., Division of Applied Research and Technology
John Whalen, M.S., M.S.A., Federal Occupational Health (Waynesville, NC)

Millie Schafer, Ph.D., served as the CDC/NIOSH Project Officer on the contract, Efficacy of Ultraviolet Irradiation in Controlling the Spread of Tuberculosis (NIOSH contract no. 200–97–2602), that formed the basis of this document. Paul Jensen, Ph.D., National Center for HIV, STD, and TB Prevention, provided frequent reviews and input into the document.

The authors wish to thank Jane Weber, Roz Kendall, Brenda J. Jones, Anne Stirnkorb, Heidi Hudson, Anne Hamilton, and Alan Lunsford for their editorial support and contributions to the design and layout of this document. Clerical support in preparing this document was provided by Diana Campbell. Web development was provided by Julie Zimmer.

Please direct comments, questions, or requests for additional information to the following:

Director, Division of Applied Research and Technology
National Institute for Occupational Safety and Health
4676 Columbia Parkway
Cincinnati, OH 45226-1998
Telephone: 1–513–533–8462 or 1–800–CDC–INFO

Special appreciation is expressed to the following individuals and organizations for providing external review or input that contributed to the development of this document:

Peter Broden
Underwriters Laboratories Inc.
Northbrook, IL

Lloyd Chapman
Philips Lighting Company
Somerset, NJ

David Dubiel
Underwriters Laboratories Inc.
Northbrook, IL

Charles E. Dunn, Sr.
Commercial Lighting Design, Inc., (Lumalier)
Memphis, TN

Melvin First, Ph.D.
Department of Environmental Health
Harvard School of Public Health
Boston, MA

Wladyslaw J. Kowalski, Ph.D., PE
Department of Architectural Engineering
The Pennsylvania State University
University Park, PA

Kevin J. Landkrohn, M.S.
Occupational Safety and Health Administration
Washington, D.C.

Shelly L. Miller, Ph.D.
Mechanical Engineering Department
University of Colorado
Boulder, CO

Edward A. Nardell, M.D.
Departments of Environmental Health and Immunology and
  Infectious Diseases
Harvard School of Public Health
Boston, MA

Paul Ninomura, PE
Indian Health Service
Seattle, WA

Nicholas Pavelchak, M.S., CIH
New York State Department of Health
Troy, NY

Stephen S. Rudnick, S.D.
Department of Environmental Health
Harvard School of Public Health
Boston, MA

Michael Soganich
Philips Lighting Company
Somerset, NJ

1 INTRODUCTION

Among infectious diseases, tuberculosis (TB) is the leading cause of death worldwide. The World Health Organization (WHO) estimates that between 2000 and 2020 almost 1 billion people worldwide will be infected with TB and, without better prevention and treatment measures, 35 million people will die during this period [WHO 2007]. In the United States, the number of new TB cases rapidly decreased for most of the 20th century. However, starting in 1985, there was a resurgence of TB cases that peaked in 1992 [CDC 2005b]. The increase was associated with a number of factors including the emergence of multidrug-resistant TB and the acquired immune deficiency syndrome (AIDS) epidemic [Burwen et al. 1995; Cantwell et al. 1994], an increase of TB in immigrants [CDC 2002], and a deterioration of public health programs designed to prevent the disease [Field 2001].

In 1994, the Centers for Disease Control and Prevention (CDC) published guidelines for preventing the transmission of *Mycobacterium tuberculosis* in healthcare facilities [CDC 1994]. These guidelines were issued to reduce the risk of transmitting TB to healthcare workers, patients, volunteers, visitors, and other persons who may be exposed in healthcare settings. An update of these guidelines has recently been completed [CDC 2005a]. The guidelines are based on development of a TB control program that depends on early identification, isolation, and treatment of persons with infectious TB. The control measures consist of administrative and environmental controls and the use of personal respiratory protection.

The CDC guidelines recommend the use of ventilation as a primary environmental control against the spread of TB in healthcare settings. However, some facilities such as homeless centers and older hospitals may not have mechanical ventilation systems. In addition, the ventilation systems in many facilities may not be designed to meet the recommended criteria and retrofitting these systems may be difficult and expensive. Because of these concerns, other environmental control methods have been suggested that may decrease exposure of healthcare workers to airborne *Mycobacterium tuberculosis*. One of these environmental control strategies is the use of ultraviolet germicidal irradiation (UVGI) in the upper portion of rooms or areas. Before publication of the 1994 CDC TB guidelines, success had been reported anecdotally from the use of upper-room UVGI systems to reduce the occurrence of TB and respiratory infections [McLean 1961; Perkins et al. 1947; Wells and Holla 1950; Wells et al. 1942; Willmon et al. 1948]. However, most of the studies were more than 30 years old when the guidelines were developed in 1994, and a lack of data existed concerning the efficacy of UVGI fixtures and lamps [Miller et al. 2002] that are designed to maximize UVGI in the upper portion of a room and minimize exposure to room occupants. Also, the results of the studies showed a variation in the effectiveness of the upper-room UVGI systems. One possible reason for the variety of outcomes noted in the studies was the lack of knowledge concerning the parameters necessary to ensure an effective system [Kowalski and Bahnfleth 2000; Riley and O'Grady 1961]. Some of the questions that were unanswered included the following:

- What is the UV fluence rate in the upper room necessary to provide a UVGI dose that will kill or inactivate an airborne surrogate of *M. tuberculosis*?

- Does the spatial distribution of UVGI influence the effectiveness of an upper-room UVGI system?

- Can a practical method be developed to accurately measure UVGI levels in the upper room and at eye level in occupied rooms and areas?

- What are the effects of ventilation, photoreactivation (PR), humidity, air mixing, and temperature on the ability of UVGI to kill or inactivate bioaerosols of a surrogate of infectious *M. tuberculosis*?

Because of the paucity of information about the effectiveness of upper-room UVGI systems, CDC's National Institute for Occupational Safety and Health (NIOSH) funded studies to determine the ability of an upper-room UVGI system in a simulated healthcare room to kill or inactivate airborne mycobacteria. These studies have shown that a properly installed and maintained system can be effective in reducing exposure to mycobacteria. A detailed technical report [Miller et al. 2002] is available on the CDC/NIOSH web site (http://www.cdc.gov/niosh/reports/contract/pdfs/ultrairrTB.pdf).

This document is intended to provide an overview of the current knowledge concerning upper-room UVGI systems and research needs. Information from the CDC/NIOSH-funded laboratory studies and other relevant studies is combined in this report to provide guidelines for the installation and use of upper-room UVGI systems. Although other pathogenic microorganisms may be killed or inactivated by upper-room UVGI systems, the guidelines were developed for the installation and use of upper-room UVGI systems capable of killing or inactivating surrogates of mycobacteria.

2 BACKGROUND

TB is a severe, contagious disease that primarily affects the lungs. It is usually transmitted from person to person via airborne *M. tuberculosis* particles known as droplet nuclei. Most droplet nuclei are believed to have an aerodynamic diameter less than 5 micrometers (μm) [Duguid 1946; Fennelly et al. 2004; Papineni and Rosenthal 1997; Wells 1955]. Because the droplet nuclei are small, they may remain airborne for long periods [Loudon et al. 1969], which increases their potential for being inhaled. Infection can occur when a person inhales the droplet nuclei that may be deposited deep in the lungs in the alveoli [Riley and O'Grady 1961]. Some of the environmental factors affecting the potential for disease transmission are the concentration of infectious particles in the air and the exposure duration [Segal-Maurer and Kalkut 1994]. Most people who become infected with TB bacteria do not develop clinical disease since their immune responses are able to arrest multiplication of the bacteria. However, some of the bacteria may remain dormant and viable in their bodies for many years [CDC 2005a]. Persons with compromised immune systems (e.g., those with AIDS, chemotherapy patients) have an increased risk of developing active disease [CDC 2005a]. It is estimated that between 5% to 10% of persons infected with *M. tuberculosis* (but not human immunodeficiency virus-infected [HIV-infected]) will develop active disease during their life [WHO 2007].

Approximately 13 million adults in the United States are infected with TB, and more than 1 million of these persons may eventually develop active disease and become capable of transmitting infection to others [62 Fed. Reg.* 54159 (1997)]. TB outbreaks have been documented in many indoor environments [CDC 2000], and more than 5 million workers in the United States may be exposed to TB in workplaces such as hospitals, homeless centers, nursing homes, and correctional facilities [62 Fed. Reg. 54159 (1997)].

The 1994 CDC TB guidelines provided recommendations to control TB based on a risk assessment process where healthcare facilities were classified according to five categories of risk and a corresponding series of administrative, environmental, and personal protective control measures [CDC 1994]. The guidelines were implemented by many healthcare facilities in the United States and helped to reduce TB outbreaks and nosocomial transmission [CDC 2002; Field 2001; Manangan et al. 1999]. In 2002, there were 15,078 new cases of TB reported in the United States, which represents a 43.5% decline in new cases from the 1992 peak [CDC 2003]. In 2003, a total of 14,871 cases were reported [CDC 2004] and in 2004, there were 14,511 new cases [CDC 2005b]. Although the number of new cases continues to decline, TB still remains a threat to healthcare and other workers [CDC 2005a; Field 2001]. At this time, an important risk to healthcare workers is from exposure to patients or persons with unsuspected, undiagnosed active TB [Field 2001]. The use of an environmental control method such as an upper-room UVGI system may be effective in reducing worker exposure to droplet nuclei containing *M. tuberculosis* [Menzies et al. 1995; Nardell 1995; Segal-Maurer and Kalkut 1994; Stead et al. 1996].

Federal Register. See Fed. Reg. in references.

2.1 UVGI Systems

The germicidal effect of solar radiation was first reported by Downes and Blunt [1877], who associated it chiefly with the actinic rays of the spectrum. The first large-scale use of a UVGI system for water purification occurred in France in 1909 or 1910, when Marseilles authorities invited proprietors of water purification apparatus to participate in competitive tests [Baker 1981]. Other early uses of UVGI systems include their use in 1936 to reduce postoperative infections [Sharp 1939], and their use in 1937 in the ventilation system of a school to reduce the incidence of measles [Wells et al. 1942]. In the late 1950s and early 1960s, Riley et al. [1957, 1959, 1962] conducted a series of animal experiments that showed conclusively that intense UVGI in air ducts kills or inactivates virulent *M. tuberculosis* in droplet nuclei.

In addition to being used to control airborne microorganisms, UVGI systems have been used recently for a variety of applications to control microbial growth in different media such as wastewater treatment facilities, air handling unit cooling coils and filter assemblies, pharmaceuticals, biohazard control, medical equipment, and food (e.g., meats) [Kowalski and Bahnfleth 2004; Rea 2000]. In healthcare facilities, three types of UVGI systems are generally used to inactivate airborne microorganisms such as *M. tuberculosis*: (1) duct irradiation, (2) UVGI lamps incorporated into room air-recirculation units, or (3) upper-room irradiation. In duct irradiation systems, one or more UVGI lamps are positioned within a duct to irradiate air being exhausted from a room or area through the duct. In room air-recirculation units containing ultraviolet (UV) lamps, a fan draws room air into the unit near lamps to disinfect the air before it is recirculated back into the room. These units may be either portable or permanently mounted. In upper-room UVGI systems, UV lamps are installed into fixtures suspended from a ceiling or mounted on a wall. The UV lamps are positioned such that air in the upper part of the room is irradiated. The intent is to maximize the levels of UV radiation in the upper part of the room and to minimize the level in the lower part of the room where occupants are located. These systems depend on good air mixing to transport the air (and thereby the microorganisms) to the upper portion of the room. Additional information about air cleaners is readily available from other sources [CDC 2005a].

2.2 UV Radiation

UV radiation is a form of electromagnetic radiation with a wavelength between the blue region of the visible spectrum and the X-ray region. For convenient classification, the International Electrotechnical Commission [CIE 1987] has divided the wavelengths between 100 mn and 400 nm into three wavelength bands: UV-A (long wavelengths, range: 315 nm to 400 nm), UV-B (midrange wavelengths, range: 280 nm to 315 nm), and UV-C (short wavelengths, range: 100 nm to 280 nm). These spectral band designations are used to define approximate spectral regions and are shorthand notations that may vary between sources [Rea 2000].

The UV-C component of solar UV radiation is filtered out before it reaches the earth. Ozone in the atmosphere reacts with UV radiation below 290 nm and prevents UV-C radiation from reaching the earth's surface [Diffey 1991]. UV-C radiation may be produced by a number of artificial sources (e.g., arc lamps, metal halide lamps). Germicidal lamps used in upper-room UVGI systems consist of low-pressure mercury vapor enclosed in special UV transmitting glass tubes. Approximately 95% of the energy from these lamps is radiated at 253.7 nm in the UV-C range [Rea 2000].

2.3 Cellular Effects

UVGI damages living cells by directly or indirectly affecting the molecular structure of nucleic acids such as deoxyribonucleic acid (DNA) [Miller et al. 1999; Setlow and Setlow 1962]. Other studies have indicated that UVGI may also affect cytoplasmic and membrane structures [Schwarz 1998].

The photobiological reaction (e.g., the formation of covalent bonds between adjacent thymine bases in DNA) that may occur when a photon of UVGI (at 254 nm) strikes a cell translates into cellular or genetic damage that may lead to cell death or inability to successfully replicate [Miller et al. 1999; Setlow 1997; Setlow and Setlow 1962]. UVGI provides a significant germicidal effect since many biological polymers absorb energy in this bandwidth [Setlow 1966].

2.4 Health Issues

In humans, UVGI may be absorbed by the outer surfaces of the eyes and skin. Short-term overexposure may result in photokeratitis (inflammation of the cornea) and/or keratoconjunctivitis (inflammation of the conjunctiva) [NIOSH 1972]. Keratoconjunctivitis may be debilitating for several days but is reversible. Because these effects usually manifest themselves 6 h to 12 h after exposure, their relationship to UVGI exposure may be overlooked. Symptoms may include an abrupt sensation of sand in the eyes, tearing, and eye pain that may be severe. Skin overexposure is similar to sunburn but does not result in tanning.

Several instances of healthcare workers overexposed to UVGI have been reported. Five workers in a hospital emergency room were reported to have developed dermatosis or photokeratitis after exposure to high UVGI levels from a germicidal lamp. An investigation of the incident determined that a UV lamp was unshielded [Brubacher and Hoffman 1996]. Additional reports of overexposure to UVGI from unshielded lamps have been reported in a hospital in Botswana [Talbot et al. 2002] and a morgue in the United States [Seitz 1992]. Other potential UVGI exposure hazards have been summarized in the CDC guidelines [CDC 2005a].

2.5 Exposure Guidelines

Occupational exposure guidelines for UVGI have been developed by CDC/NIOSH [NIOSH 1972] and the American Conference of Governmental Industrial Hygienists (ACGIH)

[2007]. The CDC/NIOSH recommended exposure limit (REL) is designed to protect workers against eye and skin injury. Detailed examples for calculating the REL at different UV wavelengths are provided in the CDC/NIOSH [NIOSH 1972] criteria document.

Based on the CDC/NIOSH REL, the maximum recommended exposure to UV is 6 mJ/cm^2 at 254 nm for a daily 8 h work shift. The ACGIH threshold limit value (TLV) at 254 nm is 6 mJ/cm^2 in an 8 h period. These recommended exposures correspond to a maximum recommended irradiance of 0.2 µW/cm^2 for 8 h exposure to UVGI at a wavelength of 254 nm. The ACGIH TLV also stipulates that these values should be used as a guide for control of exposure to continuous sources for exposure durations equal to or greater than 0.1 seconds.

Upper-room UVGI system designers have used 0.2 µW/cm^2 as the maximum lower (occupied) room irradiance [First et al. 2005; Nardell et al. 2008] to limit the irradiance level in the lower room to the 8 h REL. Some researchers [First et al. 2005; Nardell et al. 2008] believe this limits the irradiation level in the upper room thereby decreasing the potential effectiveness of the system. Many workers move around during the course of their work and may not be exposed to a single irradiance level during their work shifts [CDC 2005a; First et al. 2005; Nardell et al. 2008].

The recommended permissible exposures for various times to UVGI at 254 nm are provided in Table 1. Recommended exposures for work shifts of greater than 8 h in a 24 h period can be calculated using the formula provided in Table 1 and noted by permissible exposure times (PET). The recommended levels should not be used for photosensitive persons, persons concomitantly exposed to systemic or topical photosensitizing agents, or persons who have had the lens of the eye removed during cataract surgery. Workers exposed to UVGI above the REL require the use of personal protective clothing and equipment to protect their eyes and skin.

Table 1. Permissible exposure times for given effective irradiances at 254 nm wavelength

Permissible exposure time*		Effective irradiance ($\mu W/cm^2$)
(Units given)	(Seconds)	
8 h	28,800	0.2
4 h	14,400	0.4
2 h	7,200	0.8
1 h	3,600	1.7
30 min	1,800	3.3
15 min	900	6.7
10 min	600	10
5 min	300	20
1 min	60	100
30 s	30	200
10 s	10	600
1 s	1	6,000
0.5 s	0.5	12,000
0.1 s	0.1	60,000

*At 254 nm, the CDC/NIOSH REL is 6 mJ/cm^2 (6000 µJ/cm^2). Permissible exposure times (PET) for healthcare workers can be calculated for various irradiance levels as follows.

$$\text{PET (seconds)} = \frac{\text{REL (6000 µJ/cm}^2 \text{ at 254 nm)}}{\text{Measured irradiance level at 254 nm (µW/cm}^2\text{)}}$$

3 FACTORS INFLUENCING UVGI EFFECTIVENESS

Several laboratory-based studies have been conducted on the efficacy of upper-room UVGI systems. These studies vary in the methods used and the parameters evaluated making it difficult to directly compare results. An overview of these studies noting relevant information is provided in Tables 2 and 3.

A number of factors have been identified as potentially affecting the ability of upper-room UVGI systems to kill or inactivate mycobacteria [CDC 2005a; Miller et al. 2002]. These factors, which are described below, include UV fluence rate in the upper air, ventilation, air mixing, relative humidity (RH), photoreactivation (PR), and temperature.

3.1 UV Irradiance and Dose

Factors that must be considered when evaluating the ability of an upper-room UVGI system to kill or inactivate airborne microorganisms include the sensitivity of the microorganisms to UVGI and the dose of UVGI received by a microorganism or population of microorganisms. UVGI dose is the UV irradiance multiplied by the time of exposure and is usually expressed as $\mu W \cdot s/cm^2$. Fluence describes the incident irradiation and is sometimes used by research scientists in place of UV dose with the implicit assumption that all UV is absorbed.

As noted below, airborne microorganisms exposed to UVGI undergo an exponential decrease in population similar to that produced by ventilation and other disinfection methods. Based on research by Riley et al. [1976], it was estimated that the dose required to kill or inactivate 90% of airborne *M. tuberculosis* is 576 $\mu W \cdot s/cm^2$ [ACCP 1995]. Kowalski [2006] estimates the dose necessary to produce a 90% inactivation rate at about 1080 $\mu W \cdot s/cm^2$. Although the ability to kill or inactivate 90% of airborne *M. tuberculosis* is not necessarily a goal, it does indicate that to provide a sufficient dose to kill or inactivate a high percentage of droplet nuclei, several passes through the UVGI zone may be necessary [Beggs and Sleigh 2002; First et al. 1999b]. Because intense UVGI exposure can only occur as the bacteria move through the upper room, the fluence rate must be sufficient to kill or inactivate the bacteria during this period [ACCP 1995].

3.1.1 Sensitivity of Microbial Populations to UVGI

Several models have been developed concerning the ability of UVGI to kill or inactivate a population of microorganisms. These models are discussed in detail elsewhere [Kowalski et al. 2000]. The information presented here is intended to provide an overview of one of these models—the classical exponential decay model. This model states that microorganisms exposed to UVGI undergo an exponential decrease in population similar to that seen

Table 2. Comparison of selected upper-air UVGI room experiments*

Study	Microorganisms	Fan	Method[†]	Upper-zone irradiance (µW/cm²)	UV lamps (W)	Room size (m³)	RH (%)	Mechanical ventilation ACH	eACH	UVGI Effect Effectiveness (%)
Kethley and Branch [1972]	S. marcescens (2.7 µm CMD)	No	CG	10–170[‡]	60	46	40–50	6	39	—
Kethley and Branch [1972]	S. marcescens (5.2 µm CMD)	No	CG	10–170[‡]	60	46	40–50	6	18	—
Ko et al. [2002]	S. marcescens	No	CG	—	59[§]	46	44–64	6	—	53 (40–68)
Ko et al. [2002]	S. marcescens	No	CG	—	59[§]	46	30–52	6	—	86 (81–89)
Miller and Macher [2000]	B. subtilis spores	Yes	CG	25	15[¶]	36	—	2	—	56–58
Miller and Macher [2000]	B. subtilis spores	Yes	Decay	25	15[¶]	36	—	2	3.8	—
Xu [2001]; Xu et al. [2003]	B. subtilis spores	Yes	CG	42**	216[§]	87	50	0	—	78–82
First et al. [2007]	B. subtilis spores	Yes	CG	—	61[§]	41	50	2	3.8 (3.3–4.4)	—
First et al. [2007]	B. subtilis spores	Yes	CG	—	61[‡‡]	41	50	6	33 (30–36)	—
First et al. [2007]	B. subtilis spores	Yes	CG	—	61[‡‡]	41	50	2	49	—

*Abbreviations: ACH = air change per hour, CMD = count median diameter, eACH = equivalent air change per hour (for subject microbial species), RH = relative humidity, UV = ultraviolet, UVGI = ultraviolet germicidal irradiation.
[†]Decay method or constant-generation (CG) method (see glossary).
[‡]Applies only to the approximately 50% of the room where the UV lamps were located. No louvers were used on the fixtures.
[§]UVGI fixtures with louvers.
[¶]One 15 W UV lamp fixture without louvers.
**Spherical irradiance.
[‡‡]Louvers removed from fixtures.

Table 3. Comparison of selected upper-air UVGI room experiments using mycobacterium*

Study	Mycobacterium surrogate of virulent *M. tuberculosis*	Fan	Method[†]	Upper-zone irradiance (μW/cm²)	UV lamps (W)	Room size (m³)	RH (%)	Mechanical ventilation ACH	UVGI Effect eACH	Effectiveness (%)
Riley et al. [1976]	BCG	Yes[‡]	Decay	30[§]	46	63	20–41	0	18–33	—
Riley et al. [1976]	BCG	Yes[‡]	Decay	—	17	63	25	0	10	—
Xu et al. [2003]	BCG	Yes	CG	42 ± 19[¶]	216**	87	50	0	—	90–93
Ko et al. [2000]	BCG	No	CG	—	59**	46	41–69	6	11.7 ± 7.1	64 ± 10
Ko et al. [2000]	BCG	No	CG	—	36*	46	41–69	6–8	9.8 ± 6.4	52 ± 19
Xu [2001]	*M. parafortuitum*	Yes	CG	42 ± 19[¶]	216**	87	50	0	—	93–98
Xu [2001]	*M. parafortuitum*	Yes	CG	42 ± 19[¶]	216**	87	50	6	—	83–95
Xu et al. [2003]	*M. parafortuitum*	Yes	Decay	42 ± 19[¶]	216**	87	50	0	17.5 ± 1.8	—
Xu et al. [2003]	*M. parafortuitum*	Yes	Decay	20 ± 8.1[¶]	108**	87	50	0	6.7 ± 0.66	—
Xu et al. [2003]	*M. parafortuitum*	Yes	Decay	42 ± 19[¶]	216**	87	50	6	23.1 ± 0.78	—
Xu et al. [2003]	*M. parafortuitum*	Yes	Decay	20 ± 8.1[¶]	108**	87	50	6	14.8 ± 1.1	—

*Abbreviations: ACH = air change per hour, BCG = bacillus Calmette-Guérin [*M. bovis* BCG], eACH = equivalent air change per hour (for subject microbial species), RH = relative humidity, UV = ultraviolet, UVGI = ultraviolet germicidal irradiation.
[†]Decay method or constant-generation (CG) method (see glossary).
[‡]During aerosolization only.
[§]No upper-air UVGI measurements were taken during experiments. The figure provided is an estimate made by Nicas and Miller [1999].
[¶]Spherical irradiance.
**UVGI fixtures with louvers.

in other methods of disinfection such as occurs during exposure to biocides, heat [Riley and Nardell 1989], or radiation [Casarett 1968]. In radiation biology, this phenomenon has been described as the target theory where an event (hit) occurs in a cell that causes the death or inactivation of the microorganism [Casarett 1968]. Ko et al. [2000] assumed a one-hit, one-target model where (1) each individual microorganism in a population has the same sensitivity to UVGI, (2) a single hit by a quantum of UVGI will inactivate the microorganism, and (3) the number of hits is proportional to the UVGI dose. This model represents the effect of UVGI as a first-order rate with the logarithm of percent survival linearly proportional to the UVGI dose.

Using the one-hit model, if the relationship between the dose and surviving microorganisms is plotted on a semilogarithmic scale, the result will be a straight line that can be represented by the equation (adapted from Casarett [1968]):

$$S = e^{-kD} \text{ (or in the logarithmic form of } \ln S = -kD\text{),}$$

where S is the fraction of surviving cells,
    $-k$ is the slope of the line and is known as the decay rate constant, and
    D is the UVGI dose (irradiance × time).

Researchers have postulated that the decay rate constant (k) is directly related to the sensitivity of a microbial species to UVGI and is unique to each species [Hollaender 1943; Jensen 1964]. The rate constants have been determined in various media (e.g., air, water) to attempt to provide comparisons between different microorganisms [Brickner et al. 2003; Kowalski et al. 2000; Sylvania 1981]. Also, Kethley [1973] used this relationship to examine the sensitivity of different microorganisms and developed the Z-value. The Z-value is defined as the ratio of the inactivation rate normalized by UV irradiance and may be represented as the slope of a straight line formed from the plot of the natural logarithm of surviving microorganisms versus UVGI dose. The Z-value is virtually the same as the decay rate constant.

Due to safety concerns and difficulty with growth in the laboratory, it is not practical to aerosolize virulent *M. tuberculosis* for experimental studies. Therefore, researchers have examined bacteria to use as surrogates for *M. tuberculosis* in research studies that have a similar sensitivity to UVGI and a low degree of pathogenicity. The relative susceptibility of selected microorganisms based on their Z-values is presented graphically in Figure 1. Figure 1 is limited because relatively few Z-values have been determined for airborne microorganisms. The studies that were used to derive the figure are provided in the caption for Figure 1.

While the data in this figure are limited, it does provide a general assessment of the susceptibility of the microorganisms examined. The relatively close UVGI susceptibility of virulent *M. tuberculosis* with *Mycobacterium bovis* BCG and *Mycobacterium parafortuitum* indicates that these microorganisms may be acceptable surrogates for *M. tuberculosis*.

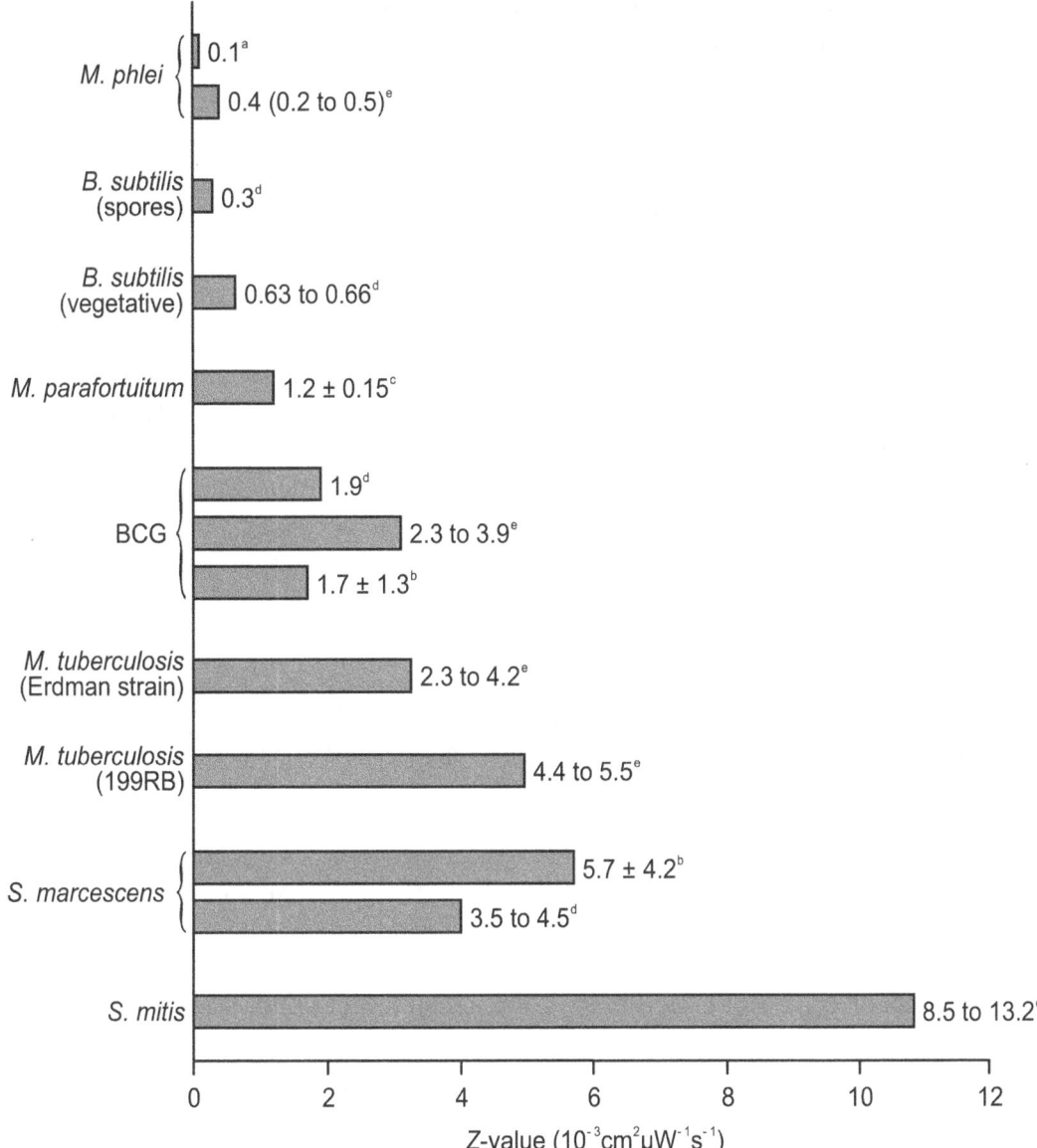

Figure 1. Relative sensitivity of selected airborne microorganisms to UVGI. The higher the Z-value, the greater the microorganisms' sensitivity to UVGI. The data sources are indicated by superscripted letters: [a]Kethley 1973; [b]Ko et al. 2000; [c]Miller et al. 2002; [d]Peccia 2000; [e]Riley et al. 1976

3.1.2 Irradiance—Experimental Results

Experimental upper-room UVGI systems used in rooms with aerosolized bacteria (including surrogates of virulent *M. tuberculosis*) have indicated that the higher the UV fluence rate produced in the upper air of a room, the greater the effectiveness of the system [Miller and Macher 2000; Miller et al. 2002; Riley et al. 1976]. In addition, as UV irradiance was increased in bench-scale reactors, there was an increase in the percentage of bacteria killed or inactivated [Peccia 2000; Riley and Kaufman 1972].

Miller et al. [2002] examined the ability of an upper-room UVGI system designed to maximize UVGI in the upper portion of a room to inactivate airborne microorganisms (including surrogates of virulent mycobacteria species) under different environmental conditions including varied UV fluence levels and lamp configurations. A picture of the simulated healthcare room used in these experiments is provided in Figure 2. The upper-room UVGI system consisted of five luminaries that were installed to provide uniform UV fluence rates based on the fixtures available. The fluence rate in the upper room at 2.3 m above the floor was determined by using a radiometer and spherical actinometry at the midpoint of the UV zone. A number of measurements were made that were evenly spaced throughout the irradiated zone and the data were spatially averaged. Based on the spherical actinometry data [Rahn et al. 1999], the system produced an average fluence rate of (42 ± 19) $\mu W/cm^2$ from the total 216 W (66 W UV-C) nominal output of the five luminaries. The following results from this study are relevant to estimating appropriate upper-room UVGI levels and lamp configurations.

- UV fluence rates below an average of 12 $\mu W/cm^2$ in the upper irradiated zone produced minimal inactivation (i.e., 1.2 equivalent air changes per hour [eACH] in a well-mixed room) for *M. parafortuitum* in this system.

- A linear dose-response relationship existed between the UVGI equivalent air-exchange rate and the average level of UV fluence rate between 12 $\mu W/cm^2$ and 42 $\mu W/cm^2$ (Figure 3).

- Four wall-mounted fixtures (72 W each) were added to the room, which increased the available wattage from 216 W to 504 W. With all fixtures activated (504 W), the average UV fluence rate in the middle of the upper irradiated zone increased from 42 $\mu W/cm^2$ to 87.2 $\mu W/cm^2$. However, the average eACH of the 504 W system only increased approximately 16% above the original 216 W system. This suggests that an upper threshold level had been reached between 216 W and 504 W in which the effectiveness of the system did not increase linearly by further increasing the total system wattage (Figure 3).

- When all of the UV lamps (216 W) were configured on a single wall, a significant UV fluence rate gradient was created with a fluence rate of 140 $\mu W/cm^2$ to 160 $\mu W/cm^2$ on the side of the room nearest the UV lamps and 2 $\mu W/cm^2$ to 10 $\mu W/cm^2$ on the opposite side. Although the average UV fluence rate (43.5 $\mu W/cm^2$) for the entire upper portion of the room was nearly identical to the original lamp configuration with more uniform UV distribution, the UVGI eACH decreased by approximately 30% when using the unbalanced configuration.

Figure 2. Simulated healthcare UVGI room, University of Colorado. (Courtesy of Professor Shelly L. Miller, University of Colorado, Boulder, Colorado).

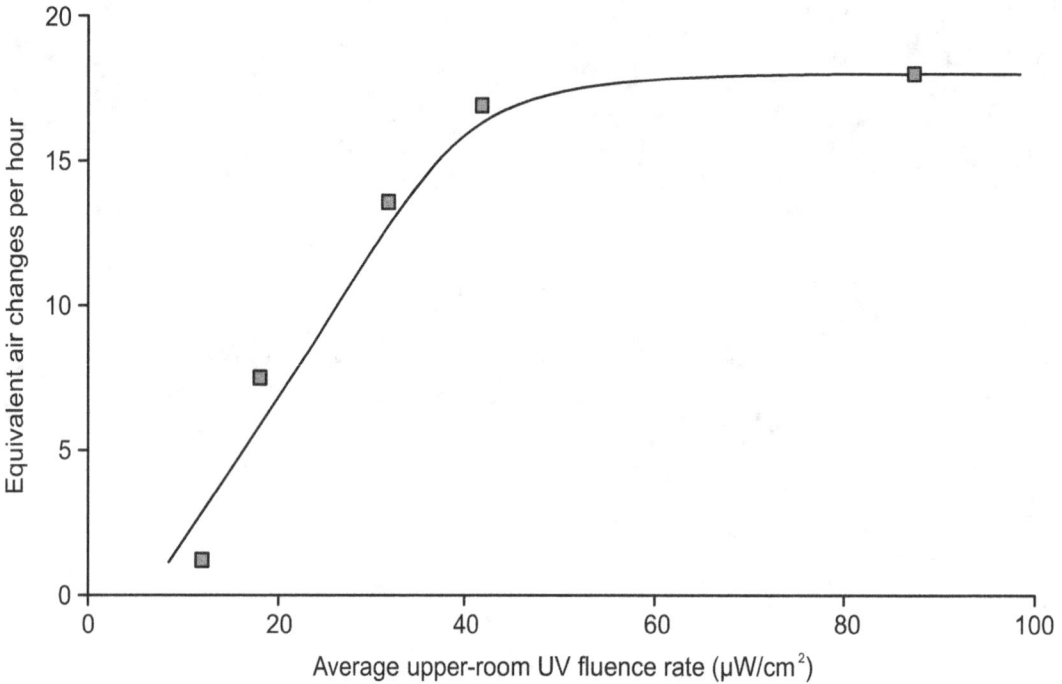

Figure 3. UVGI-induced inactiviation of *M.parafortuitum* in a test room under well-mixed conditions at 50% relative humidity. Adapted from Miller et al. [2002].

- In the lower zone at 1.5 m above the floor, the average UV fluence rate was 0.2 µW/cm^2 (standard deviation [SD] = 0.04 µW/cm^2) when the original 216 W UVGI system was activated. This value is at the maximum occupational irradiance level of continuous exposure for an 8 h day recommended by CDC/NIOSH for UV-C radiation at 254 nm [NIOSH 1972].

3.1.3 Basis for the UVGI Irradiance and Lamp Configuration Guidelines

Based on the results of room experiments with upper-room UVGI systems (see Tables 1 and 2) and aerosolized bacteria in bench-scale reactors [Peccia 2000; Riley and Kaufman 1972], it is apparent that the greater the UV fluence rate in the irradiated zone, the more effective the system. However, there appears to be an upper threshold where an increase in UVGI does not directly correspond to an increase in the ability of the system to kill or inactivate microorganisms [Beggs and Sleigh 2002; Lidwell 1994; Memarzadeh 2000; Miller et al. 2002].

It is difficult to determine the most effective upper-room UV fluence rate based on previously published room studies (see Tables 2 and 3). However, the comprehensive study by Miller et al. [2002] indicates that an upper-room UVGI system designed to provide an average UV fluence rate in the range of 30 µW/cm^2 to 50 µW/cm^2 may be effective in killing or inactivating *M. tuberculosis*. Although not directly comparable, this is consistent with the results for *M. bovis* BCG obtained by Riley et al. [1976] in a room where an average of 25 eACH was obtained with an upper-room system that produced an average UV fluence rate estimated to be approximately 30 µW/cm^2 [Nicas and Miller 1999]. Systems that provide a lower irradiance may still be effective, but to a lesser degree, in killing or inactivating airborne *M. tuberculosis*. Systems that are designed to provide higher fluence rates may not be cost effective since they may not significantly increase the ability of the upper-room system to kill or inactivate *M. tuberculosis*.

The results of the Miller et al. [2002] study indicate a decrease in the effectiveness of the UVGI system when the UV fixtures were placed on only one side of the room. This is consistent with the findings of Riley and Permutt [1971] where their results indicated that a wider distribution of low-irradiance UV lamps was more efficient when compared with the use of one centrally located high-irradiance UV lamp. This suggests that upper-room UVGI systems should be installed to provide as uniform a UVGI distribution in the upper air as possible.

It should be noted that potentially infectious droplet nuclei emitted from an infected host may be coated with mucus and consist of more than one bacterium. The bacteria in most of the studies that form the basis of the irradiance guidelines provided in this document were primarily single cells that were aerosolized in deionized water. This lack of a mucus coating may result in an increased sensitivity to UVGI when compared to bacteria in droplet nuclei from an infected host [Lai et al. 2004].

> **UVGI Irradiance and Lamp Configuration Guidelines**
>
> Based on experimental studies, a well-designed upper-room UVGI system may be effective in killing or inactivating most airborne droplet nuclei containing *M. tuberculosis* if designed to provide an average UVGI irradiance in the range of 30 µW/cm^2 to 50 µW/cm^2 provided the other elements stipulated in these guidelines are met. In addition, the fixtures should be installed to provide as uniform a UVGI distribution in the upper room as possible.

3.2 Upper-Room UVGI Systems and Mechanical Ventilation

Ventilation rates for healthcare facilities are often expressed in the number of air changes per hour (ACH). One ACH means that an equivalent amount of the total volume of air in a room will be replaced in the room in 1 h. Under ideal conditions in a room where droplet nuclei are released at a single point in time, mechanical room ventilation reduces the number of droplet nuclei in the room in a logarithmic fashion when plotted against time. Under these conditions, for one air change, the original number of droplet nuclei would be reduced by 63%, leaving 37% in the room [Mutchler 1973]. After two air changes, 14% of the contaminant remains (37% × 37%) and after three air changes, about 5% (37% × 14%) of the contaminant remains. Table 4 provides the particle removal efficiency of a perfectly mixed room at 1 h for various ACH rates. In this table, it is assumed that new particles are not being introduced into the room beyond the initial starting time (see decay model in glossary).

Table 4. Particle removal efficiency for various rates of air change per hour (ACH) in a perfectly mixed room ($K = 1$)

ACH	Removal efficiency at one hour (%)*
1	63.2
2	86.5
3	95.0
4	98.2
6	99.75
12	99.9994
20	99.99999

*Given by $100 \times (1 - 0.368^A)$, where A = air change per hour.

The American Institute of Architects (AIA) has developed recommended ventilation requirements for areas affecting patient care in hospitals and outpatient facilities [AIA 2006]. Their recommendations include minimum ventilation rates of 2 ACH in patient corridors, 6 ACH in patient rooms, 12 ACH in airborne infection isolation rooms, 12 ACH in protective environment rooms, and 12 ACH in bronchoscopy and emergency room waiting areas. The recommendations were developed for comfort, asepsis, and odor control. Additional recommendations for designing and operating ventilation systems have been published [ACGIH 2004; ASHRAE 2007]. Ventilation engineers in collaboration with infection control and occupational health staff should design ventilation systems for healthcare facilities [CDC 2005a].

As noted above, mechanical ventilation removes contaminants from a room in an exponential fashion [ACCP 1995; Mutchler 1973; Nardell et al. 1991]. Also, the action of UVGI in killing or inactivating microorganisms in a room approximates an exponential curve [Kowalski and Bahnfleth 2000]. When 63% of droplet nuclei in a room are killed or inactivated by UVGI, this is equivalent to one ACH in terms of reducing the total droplet nuclei concentration in the room [First et al. 1999b]. This reduction of droplet nuclei in a room by a method other than mechanical ventilation (e.g., upper-room UVGI systems or portable air cleaners) is referred to in this document as equivalent ACH or eACH. As noted previously, the sensitivity of different microbial species to UVGI varies. Therefore, the eACH for upper-room UVGI systems will be different for each microbial species. For example, the eACH obtained for *M. tuberculosis* will be different from that of other microbial species such as *Serratia marcescens*. Also, due to experimental variance in room studies of upper-room UVGI systems, it is difficult to compare the eACH of a system between different studies.

As the mechanical ventilation rate in a room is increased, the total number of microorganisms removed from the room via this system is increased [Memarzadeh 2000]. However, when mechanical ventilation is increased in a room where an upper-room UVGI system has been deployed, the effectiveness of the UVGI system may be reduced because the residence time of the bacteria in the irradiated zone decreases [Collins 1971; First et al. 2007; Kethley and Branch 1972; Xu 2001; Xu et al. 2003].

First et al. [2007] found a lower reduction of cultural microorganisms (*S. marcescens* and *B. subtilis* spores) at 6 ACH than at 2 ACH in seven of eight pairs tested. The fraction of cultural microorganisms remaining at 2 ACH was 26% compared to 75% remaining at 6 ACH. Although the effectiveness of an upper-room UVGI system may be reduced by increasing the mechanical ventilation, other experimental-room studies have not indicated a significant reduction in effectiveness at mechanical ventilation up to 6 ACH. During a study in a simulated healthcare room, Xu et al. [2003] used aerosolized *M. parafortuitum* in decay-method experiments that were conducted at mechanical ventilation rates of 0, 3, and 6 ACH, with a mixing fan to ensure a well-mixed room. The upper-room UVGI system (216 W) provided (42 ± 19) $\mu W/cm^2$ in the irradiated zone as measured by chemical actinometry. Inactivation rates (eACH) provided by the UVGI system were 17.5 ± 1.8 at 0 ACH, 20.0 ± 2.4 at 3 ACH, and 23.1 ± 0.78 at 6 ACH. Constant generation (CG) method experiments using *Bacillus subtilis* spores at 0 and 6 ACH (with mixing fans) noted a decrease in effectiveness at 6 ACH. However, possibly due to the limited number of experiments that were conducted, the decrease was not statistically significant. In CG method experiments for *M. parafortuitum*, increasing the ventilation rate from 0 ACH to 6 ACH did not statistically decrease the effectiveness. Ko et al. [2002] used bioaerosols of *S. marcescens* and *M. bovis* BCG under CG conditions to examine the efficacy of a UVGI system (nominal 59 W) consisting of ceiling- and wall-mounted fixtures. Without the use of a mixing fan, the UVGI effectiveness for *S. marcescens* at a mechanical ventilation of 2 ACH was 46% and at 6 ACH was 53%. Using a mixing fan, the UVGI effectiveness for *S. marcescens* was 62% at 2 ACH and 86% at 6 ACH. At 6 ACH, without a mixing fan, the UVGI killed or inactivated 64% of the BCG aerosols. Using *S. marcescens*, Kethley and Branch [1972] reported an eACH of up to 39 with a ventilation rate of 6 ACH. Studies by Miller et al. [2002] and Kujundzic et al. [2006] examined the relationship between an upper-room UVGI system, portable air cleaners, and ventilation rates of 0 and 6 ACH and found that, as long as the air was well mixed, the particle removal rates of the three systems were additive.

> **Upper-Room UVGI Systems and Mechanical Ventilation Guidelines**
>
> Under experimental laboratory conditions, the rate that microorganisms are killed or inactivated by UVGI systems appears to be additive with mechanical ventilation systems in well-mixed rooms (up to mechanical ventilation rates of 6 ACH).

3.3 Air Mixing

Upper-room UVGI systems rely on air movement between the lower portion of the room where droplet nuclei are generated and the upper irradiated portion of the room. Once in the upper portion, droplet nuclei containing *M. tuberculosis* may be exposed to a sufficient dose of UVGI to kill or inactivate them.

The efficacy of upper-room UVGI systems is better in well-mixed rooms than in poorly mixed rooms [First et al. 2007; Xu et al. 2000]. Although 100% vertical air mixing between the lower and upper portions of a room is probably impossible to achieve, in most rooms the air is sufficiently well mixed [Beggs and Sleigh 2002].

A number of factors including the following may affect the vertical mixing of air between the upper irradiated portion of the room and lower occupied portion of the room:

- Temperature differential between the supply air and room air.
- Location of the supply diffusers and exhaust grills.
- Mechanical ventilation rate.
- Location of furniture and movement of people.
- Changing environmental conditions (e.g., winter or summer conditions of heating or cooling).
- Use of fans.

3.3.1 Temperature and Mechanical Ventilation Considerations

The ventilation regime and temperature of the supply air (i.e., winter or summer conditions) can affect room air mixing. Several studies have examined the effect of air temperature on air mixing and UVGI efficacy [Miller et al. 2002; Riley and Kaufman 1971; Riley and Permutt 1971]. These studies indicated that the efficacy of UVGI was greatly increased if cold supply air relative to the lower portion of the room entered through diffusers in the ceiling (which may happen in the summer when the air-conditioning system is functioning). When warm air entered the upper portion of the room (such as might happen during the winter when the heating system is functioning), mixing between the warm air and the cold air in the lower portion of the room was minimal. The reports concluded that temperature gradients (e.g., 3 °F or more as noted by Riley et al. [1971]) between the upper and lower portions of the room favored (i.e., cold air entering through supply diffusers in the upper portion of the room) or inhibited (i.e., hot air entering through supply diffusers in the upper portion) vertical mixing of air between the two zones. It is important to note that the ASHRAE handbook (fundamentals) indicates when vertical air temperature differences of 4 °C exist between the upper and lower portions of a room, approximately 10% of the population are not comfortable, and as the temperature gradient increases, the individual dissatisfaction increases [ASHRAE 2005].

Ko et al. [2002] noted a 16% increase in UVGI effectiveness at a ventilation rate of 2 ACH and 33% at 6 ACH for aerosolized *S. marcescens* when a portable mixing fan was used. Xu et al. [2000, 2003] used a simulated healthcare room to examine the efficacy of upper-room UVGI under incomplete mixing conditions with *M. parafortuitum* as the test microorganism. Under the experimental conditions, the UVGI efficacy decreased slightly (97% to 90%) without the use of a mixing fan during summer conditions with cold air entering from a supply diffuser in the upper portion of the room. However, at a ventilation rate of 6 ACH during winter conditions (i.e., the upper portion of the room was 29 °C and the lower portion 25 °C due to hot air entering through the supply diffuser in the upper portion of the room), the UVGI efficacy decreased from 89% to 19% without the use of a portable box fan.

3.3.2 Placement of Supply Diffusers and Exhaust Grills

Noakes et al. [2004] used a two-zone ventilation model to examine the effect of air mixing on upper-room UVGI systems. The upper zone represented the UVGI field and the lower zone represented the occupied portion of the room. Four ventilation scenarios were examined where the supply and exhaust vents were on opposite walls near the ceiling or near the floor. In this model, the worst configuration for air mixing occurred when both the supply and exhaust vents were in the upper zone. The best air-mixing configuration was represented by having both the supply diffusers and exhaust grills near the floor. The two other scenarios examined (supply upper zone and exhaust lower zone; supply lower zone and exhaust upper zone) provided good air mixing similar to the scenario where both the supply and exhaust vents were in the lower zone. Temperature differentials between the supply air and the room air were not considered in this model.

Xu and Miller [1999] used tracer gas (SF_6 in He) released in the lower portion of a simulated healthcare room to examine air mixing. Experiments were conducted with the mechanical ventilation at 0, 3, or 6 ACH; mixing fans on or off; and the temperature of the supply air varied (two supply diffusers were on one side of the ceiling and two exhaust grills were on the other side of the ceiling). The mixing between the upper and lower portions of the room was relatively good under all conditions examined except during winter conditions when the supply air (32 °C) was warmer than the room air (21 °C) and the mixing fans were off. Under these conditions, the concentration of tracer gas was approximately 20 times higher in the lower zone than the upper zone, indicating poor mixing. This suggests that if supply diffusers and exhaust grills are both located in the ceiling, incomplete mixing may occur under wintertime conditions. The buoyancy of the warmer supply air (warmer than room air) causes a short circuit between the supply diffuser and exhaust grill [Miller et al. 2002]. It is important to recognize there are many different types of room air diffusion systems. Air velocities exiting the diffuser are generally much higher than what would be experienced in the occupied zone. When diffusers are selected for this type of application, the user should consider the drop (vertical distance of projected airstream) and the throw (horizontal distance of projected airstream) of the diffuser and how that may impact vertical mixing within the room.

3.3.3 Evaluating Air Mixing

Air mixing can be evaluated using various methods. A qualitative measure of air mixing is the visualization of air movement by generating chemical smoke using agents such as titanium tetrachloride in several areas of the room [CDC 2005a]. Smoke tubes, smoke candles, and smoke bombs are available for studying airflow patterns. Smoke movement in all areas of a room indicates good mixing. Whenever such devices are used, care should be taken to avoid potential fire hazards and prevent the fire alarm system from going off in accordance with instructions from the manufacturer. In addition, caution should be used to prevent inhalation of chemical smoke, particularly by patients with respiratory conditions [Jensen et al. 1998].

A more quantitative method to determine effective mixing is through the use of tracer gas methods. Common tracer gases include sulfur hexafluoride, nitrous oxide, carbon dioxide, and bromotrifluoromethane [Persily 1988]. Tracer gas experiments should only be performed by trained persons.

> **Air Mixing Guidelines**
>
> When upper-room UVGI systems are installed, general ventilation systems should be designed to provide optimal airflow patterns (e.g., as close to $K = 1$ as possible) within rooms and prevent air stagnation or short-circuiting of air from the supply diffusers to the exhaust grills. Also, heating and cooling seasons should be considered and the system designed to provide optimal convective air movement. If the upper portion of a room is about 3 °F (1.7 °C) colder than the lower portion of the room; vertical air mixing between the zones is favored without causing significant discomfort to

> room occupants. If the upper room is warmer than the lower room, vertical air mixing is inhibited.
>
> Most rooms or areas with properly installed supply diffusers and exhaust grills should have adequate mixing. If areas of air stagnation are present, air mixing should be improved by adding a fan or repositioning the supply diffusers and/or exhaust grills. **If there is any question about vertical air mixing between the lower and upper portions of the room due to environmental or other factors, a fan(s) (e.g., axial box fan) should be used to continually mix the air.**

3.4 Humidity

Several studies have examined the relationship between ambient RH and the effects of UVGI on airborne microorganisms. Two studies [Rentschler and Nagy 1942; Rentschler et al. 1941] reported that RH did not have any effect on the ability of UVGI to kill or inactivate *Esherichia coli*. However, most reports [Gates 1929a; Ko et al. 2000; Luckiesh and Holladay 1942; Miller et al. 2002; Peccia 2000; Riley and Kaufman 1972; Wells 1942, 1955; Wells and Wells 1936] have indicated that UVGI effectiveness decreases as RH increases. Although not necessarily relevant to UVGI effectiveness, other studies [Peccia and Hernandez 2001; Xu 2001] have shown that as RH increases, the ability of bacteria to repair UVGI damage to their DNA via PR increases.

Contradictory results have been reported for lower RH levels. That is, there is also an indication that UVGI effectiveness decreases at lower RH levels (e.g., 25%) when compared with UVGI effectiveness at 50% RH, particularly if the UVGI dose is relatively low (e.g., 1 $\mu W \cdot s/cm^2$) [Miller et al. 2002; Peccia et al. 2001; Riley and Kaufman 1972]. Additional research needs to be done at lower RH levels to examine the potential for decreasing UVGI effectiveness [Miller et al. 2002].

Ko et al. [2000] provided several theories for the protection from UVGI apparently provided to microorganisms by high RH levels. Peccia [2000] suggested that the DNA conformation (based on hydration state) and associated photochemical damage may, at least in part, be responsible for the decrease in UVGI effectiveness on airborne bacteria at higher RH. While the reason for the decrease in UVGI effectiveness as RH increases is not clearly understood, RH needs to be considered in the general context of upper-room UVGI systems.

In a simulated healthcare room study under well-mixed conditions at a ventilation rate of 6 ACH, Miller et al. [2002] varied the RH from 25% to 100% and examined the effectiveness of their UVGI system (216 W) to kill or inactivate airborne *M. parafortuitum*. When compared with 50% RH, the effectiveness of the system decreased by approximately 40% for both 75% RH and 100% RH. At 25% RH, the effectiveness was reduced by approximately 20% when compared with the effectiveness at 50% RH.

Wells and Wells [1936] reported a sharp decline in the ability of UVGI to kill or inactivate airborne *E. coli* in a small chamber when the humidity increased from 55% RH to 65% RH. Riley and Kaufman [1972] examined the effects of RH and UVGI irradiation on airborne *S. marcescens*. They used a bench-scale apparatus where the microorganisms were exposed to UVGI doses from a UVGI lamp (from 0.75 µW·s/cm^2 to 96 µW·s/cm^2) at different humidity levels (25% RH to 90% RH) as they passed through a bank of parallel glass tubes. The research indicated a sharp decline in the ability of UVGI to inactivate *S. marcescens* at RH values above 60% to 70%. Above 80% RH, there was evidence for reactivation of the bacteria. Ko et al. [2000] used an experimental bench chamber to evaluate the effect of RH and UVGI on aerosols of *M. bovis* (BCG) and *S. marcescens*. The microorganisms were subjected to UVGI doses from 57 µW·s/cm^2 to 829 µW·s/cm^2 at three different RH levels (approximately 30%, 60%, and 90%). Increased humidity resulted in a lowered susceptibility of the microorganisms to being killed or inactivated by UVGI and when the RH was above 85%, an increase occurred in resistance of both *S. marcescens* and *M. bovis* BCG to UVGI inactivation. The *Z*-values decreased about 50% at an RH of about 85% when compared to an RH of around 60%. Peccia et al. [2001] conducted aerosol experiments using a bench-scale reactor (0.8 m^3) to examine the relationship of RH on *M. parafortuitum*, *S. marcescens*, and *B. subtilis* endospores and vegetative cells at different RH levels between 20% and 95%. There was a significant reduction in *Z*-values that indicated a decrease in the ability of this UVGI system to kill or inactivate the bacteria as the RH increased. The maximum *Z*-value was seen at an RH near 50% and the minimum *Z*-value at an RH of 95% for all three microorganisms studied. Over this RH range, the *Z*-value for *S. marcescens* decreased by a factor of 6.5, *B. subtilis* by a factor of 1.5, and *M. parafortuitum* by a factor of 3.2.

3.4.1 Photoreactivation and Relative Humidity

Photoreactivation (PR) is the ability of some microorganisms to repair UVGI damage to their DNA (see *Photoreactivation* in the glossary). Exposure to visible light (380 nm to 430 nm) is a component for one type of photoreaction repair mechanism. Since fluorescent and incandescent light and sunlight can be expected to be present where upper-room UVGI systems are in use, the ability of the systems to inactivate airborne mycobacteria may be affected by PR. David et al. [1971] showed that *M. tuberculosis* H37Ra and *Mycobacterium marinum* were capable of undergoing PR. However, the cells were in liquid media when they were irradiated in Petri plates, and therefore a direct comparison cannot be made with bioaerosols [Peccia 2000].

Miller et al. [2002] used a simulated hospital room to test for PR on aerosolized *M. parafortuitum* at RH levels of 40% and 100%. Only two experiments were conducted and the results of this room study were inconclusive about whether PR occurred.

Peccia and Hernandez [2001], using a bench-scale reactor, observed that aerosolized *M. parafortuitum* cells underwent PR (as indicated by approximately a twofold decrease in UVGI effectiveness between "light" and "dark" experiments) at 80% RH. When the UVGI dose was held constant and the RH increased, the rate of PR increased. The PR response was also dependent on the UVGI dose. As the dose increased (with increasing irradiance), the amount of PR decreased and at the highest UVGI irradiance (7.53 $\mu W/cm^2$) tested (the authors indicate this is equivalent to approximately 35 $\mu W/cm^2$ for the upper level of a full-scale room), no PR was observed even at high RH levels.

3.4.2 Basis for Humidity Guidelines

Room RH studies [Ko et al. 2002; Miller et al. 2002; Riley and Permutt 1971] indicated little if any effect on UVGI efficacy at RH levels up to approximately 60%. As RH increases above 60%, the UVGI levels needed to kill or inactivate airborne bacteria increase and may approach that needed to kill or inactivate microorganisms in liquid [Peccia 2000]. In a study by Peccia and Hernandez [2001], PR of bioaerosols of *M. parafortuitum* did not occur at an RH of 65% or below; however, a significant PR effect existed at 80% RH. The reason for the decrease in UV effectiveness and the increase of PR as RH increases above 60% is not clearly understood. However, the effects need to be considered in the general context of upper-room UVGI systems. Based on this, to optimize effectiveness of the system, the humidity should be controlled to 60% RH or less if upper-room UVGI systems are installed. This is consistent with most current recommendations for providing acceptable indoor air quality [AIA 2006; ASHRAE 2004, 2007].

Many of the RH studies were done using small-scale reactors instead of full-scale test rooms. Therefore, the humidity effect may have been greater than that found in real-world situations due to the difference in test conditions. However, systems that are used at humidity levels above 60% RH may still be effective in killing or inactivating airborne mycobacteria.

Humidity Guidelines

For optimal efficiency, RH should be controlled to 60% or less if upper-room UVGI systems are installed. This is consistent with the AIA [2006] and the ASHRAE [2007] recommendations that the RH affecting patient care areas in hospitals and outpatient facilities be maintained within a range from 30% RH to 60% RH. If high humidity conditions are present, increased UV irradiation levels may be necessary to achieve equivalent effectiveness.

3.5 Temperature

A relatively high or low temperature in the room or area where an upper-room UVGI system is located may affect the efficiency of the system. First, high or low temperatures may decrease the UVGI output of low-pressure mercury lamps or decrease the microorganisms' sensitivity to UVGI. Second, the old (but still commonly in use) mechanical/magnetic ballasts may be affected by high or low temperatures.

A relationship exists between lamp operating temperature and output in all low-pressure mercury lamps. The UV efficiency of the lamp is directly related to the (saturated) mercury pressure, which depends on the spot with the lowest temperature on the lamp [Philips 2006]. UVGI lamps have an optimal operating point based on temperature and airflow [Rea 2000]. UV output is standardized at this temperature and higher or lower temperatures may decrease the lamp output [VanOsdell and Foarde 2002]. For example, Philips [1992] states that the maximum output at 254 nm is produced in still air at 20 °C (68 °F). At 10 °C (50 °F), the output at 254 nm is reduced to about 88% of the maximum output. Westinghouse [1982] also notes that the output of their lamps is reduced when the ambient temperature is above or below their design temperature. The Westinghouse [1982] product literature states that the design temperature of their lamps is 27 °C (80 °F) and that the output is reduced by approximately one third at 5 °C (40 °F). The decrease in UV output will be even greater if there is airflow past the lamp due to a more rapid heat loss of the mercury vapor in the lamp [First et al. 1999a]; Sylvania [1981] and Philips [2006] also note that low temperatures reduce the operating life of their lamps.

Gates [1929b] and Rentschler et al. [1941] conducted early studies on the effect of temperature on microorganisms on Petri plates. Both studies indicated that the temperature (5 °C to 37 °C) at which the bacteria were initially exposed did not influence UVGI effectiveness. Ko et al. [2002] studied the effect of temperatures from 4 °C to 25 °C on aerosols of *S. marcescens* and *M. bovis* BCG in a chamber and noted a twofold to threefold decrease in effectiveness of the UVGI system as the temperature decreased. The chamber, including the UV lamps, reflected the outside temperature at the time of each experiment. At lower temperatures (i.e., 4 °C), this may have caused a decrease in the UVGI output of the lamps [Philips 2006] resulting in a decrease in efficiency of the UVGI system. Also, the reduced efficacy of upper-room UVGI at lower temperatures may be due to a reduced sensitivity of microorganisms at lower temperatures or a combination of lower UV output and reduced sensitivity of the microorganisms (Ko et al. 2002).

> **Temperature Guidelines**
>
> ASHRAE [2007] and AIA [2006] recommend that the design temperature for most areas affecting patient care in hospitals and outpatient facilities range from 68 °F to 75 °F (20 °C to 24 °C). This temperature range is consistent with use of low-pressure mercury lamps.

4 PRACTICAL GUIDELINES FOR INSTALLING UPPER-ROOM UVGI SYSTEMS

Upper-room UVGI systems are an adjunct to other ventilation measures and may be used in isolation rooms, areas or rooms where patients with active TB are located, and rooms where high-risk procedures are performed (e.g., bronchoscopy, sputum induction, administration of aerosolized medications). UVGI systems may also be used in patient rooms, waiting rooms, emergency departments, corridors, central areas, medical settings in correctional facilities, and other large rooms or areas where persons with undiagnosed TB could potentially contaminate the air [CDC 2005a].

4.1 UV Lamps

The most common way to generate germicidal UV radiation in lamps used in well-designed, upper-room UVGI systems is to pass an electrical charge through low-pressure mercury vapor that has been enclosed in selected glass tubes that transmit only certain UV wavelengths. This is referred to as a germicidal lamp because almost 95% of the UV radiation produced is near 254 nm in the UV-C range [Rea 2000]. The arc produced in a typical fluorescent lamp operates in a similar manner with the only difference being that the glass in the fluorescent bulb is coated with phosphor that converts UV radiation to visible light. The coated glass does not transmit germicidal wavelengths in fluorescent lamps. An average of about 35% of input watts is converted to UV-C W [Philips 2006].

Care must be used in selecting the correct UVGI lamp for use in upper-room UVGI systems. UV lamps are made for a variety of purposes that have a negligible effect in killing or inactivating airborne microorganisms. Some UV lamps (such as those used for tanning) radiate energy in the UV-A and/or UV-B range and over extended periods may have adverse health consequences for exposed persons. Other UV lamps are designed to emit radiation at 184.9 nm and produce ozone, which is hazardous to humans even at low concentrations. Low-pressure mercury lamps should be rated for low or no ozone generation [First et al. 1999a].

The germicidal activity of medium-pressure mercury lamps and pulsed-UV lamps is being examined for use in water treatment facilities [Zimmer and Slawson 2002]. However, the potential use of these lamps in upper-room UVGI systems has not been evaluated. Therefore, only low-pressure mercury lamps that are made as germicidal lamps for use in upper-room UVGI fixtures should be used at this time. The various types of low-pressure mercury UVGI lamps commercially available are listed in the latest edition of the Illuminating Engineering Society of North America (IESNA) handbook [Rea 2000].

4.1.1 Other Considerations

UV lamps are potentially hazardous since they emit UV-C radiation, contain mercury, and may cause cuts or lacerations if broken. Therefore, in accordance with 29 CFR[†] 1910.1200, material safety data sheets (MSDS) should be requested from the lamp manufacturer and be readily available to workers. The MSDS should provide information about the hazards associated with UV lamps, health effects, and precautions for safe handling and disposal.

If a lamp is broken at the worksite, at a minimum, hand and eye protection should be used for clean up. Clean up should only be performed by trained workers. The waste from the broken lamp should be disposed of as hazardous waste in the same manner as that indicated below for lamp disposal. To reduce potential mercury exposure to persons near broken lamps and since all lamps must eventually be discarded, each lamp should contain only a relatively low amount of mercury (i.e., 5 mg or less).

UV lamps require a ballast to operate. The ballast provides a high initial voltage to initiate the discharge and then quickly limits the lamp current to sustain the discharge safely. Most lamp manufacturers recommend one or more ballasts to operate their lamps. The ballasts recommended by the manufacturer should be used for each lamp type [Philips 2006; VanOsdell and Foarde 2002].

UV lamp ballasts that cause harmonic distortion may affect sensitive electronic equipment in healthcare facilities. Therefore, new or replacement ballasts should be solid state electronic and have a total harmonic distortion of less than 10% and comply with all Federal Communications Commission (FCC) rules and regulations, Title 47 CFR Part 18 for nonconsumer equipment. Electronic preheat ballasts provide the proper conditions for long lamp life, especially if the lamps are switched off and on frequently. The average lamp life will be longer if a UVGI system is only used intermittently. [Philips 2006].

4.1.2 Lamp Life

Several manufacturers of UV lamps consider 8,000 h or 9,000 h to be the effective lamp life for UVGI lamps made for upper-room systems [GE 2005; Osram 2005; Philips 2006]. The average effective life of a lamp decreases the more frequently it is turned on/off and may decrease relative to the difference between the ambient temperature of the lamp and the temperature at which the rated average lamp life was determined [Philips 2006].

Periodic replacement (e.g., on a yearly basis) of all (or a group) of UV lamps at one time may be more cost effective than spot replacement considering the time spent by maintenance personnel. Therefore, for many upper-room UVGI systems, group lamp replacement once a year would help to ensure an effective level of UV irradiation and may be cost effective.

[†]*Code of Federal Regulations.* See CFR in references.

4.1.3 Disposal

All applicable local, State, and Federal regulations must be followed when the lamps are discarded. In accordance with 40 CFR Part 273, lamps containing mercury are considered a hazardous waste and may be subject to the Resource Conservation and Recovery Act (RCRA) hazardous waste requirements. If large numbers of the lamps are broken during disposal, appropriate personal protective equipment such as gloves, safety glasses, and respiratory protection may be necessary. Whenever respiratory protection is required, it should be used in the context of a complete program as outlined in the Occupational Safety and Health Administration (OSHA) Respiratory Protection Standard [29 CFR 1910.134]. A respiratory protection program must include a written program, fit testing, medical clearance, and training (see 29 CFR 1910.134 for a complete list of program elements). OSHA offers expert assistance on the proper selection of respiratory protection and the development of change schedules for gas/vapor cartridges (http://www.osha.gov/SLTC/etools/respiratory/index.html). All disposal options should be evaluated.

4.2 UVGI Fixtures

Early UVGI fixtures included designs where UV lamps were placed in sheet metal housings to deflect UVGI into the upper portion of the room [Dumyahn and First 1999; Luckiesh 1946; Wells 1955]. These fixtures were mounted on the ceiling or hung on walls. Some UVGI lamp systems use fixtures with louvers (i.e., parallel plates) to produce a quasi-collimated narrow UVGI beam that is directed upwards at a small angle (e.g., 3° to 5° as recommended by Nardell and Riley [1992]).

In a well-designed, upper-room UVGI system, the irradiance level in the upper air increased two to four times when the louvers were removed [Miller and Macher 2000; Miller et al. 2002]. However, unshielded lamps (i.e., without louvers) should be used only in areas that are not occupied and safety features are installed to ensure that overexposure to UVGI cannot occur. Although louvers decrease the irradiance level in the upper room provided by the lamps, they also reduce exposure of room occupants to UVGI. These fixtures can be hung in the middle of a room (pendant-type) or attached to walls or in corners. They generally have parallel louvers that are coated with a nonreflective material and are designed to be used in rooms with ceilings as low as 2.4 m (8 ft). Fixtures that are designed to be used in rooms with higher ceilings (e.g., 2.7 m [9 ft]) may be used without louvers. These fixtures have upward facing flanges (baffles), deflect UVGI upward, and generally, as noted above, provide a higher irradiance level than fixtures with louvers that contain the same number and type of UV lamps. Since this type of fixture radiates UVGI upward, particular attention should be paid to potential reflection off the ceiling and other reflective surfaces. Caution should be used if UVGI upper-room systems are installed in rooms with low ceilings (i.e., less than 2.4 m [8 ft]) due to the potential exposure of room occupants to

UVGI [Dumyahn and First 1999]. Several fixtures used in upper-room UVGI systems are shown in Figure 4. Figure 4A shows pendulum and wall-mount louvered units made from stainless steel. These units are 61 cm (24 in.) in length and 20 cm (7.9 in.) in height. The unit comes in two widths (11.4 cm [4.5 in.] or 22.9 cm [9 in.]) and has a nominal output of either 25 W (8.5 W UV-C) or 50 W (17 W UV-C). The rated average effective life of the UV lamps used in these units is 8,000 h. Figure 4B shows a ceiling pendant fixture made from aluminum. It is 30.5 cm (12 in.) in height, and has a diameter of 45.7 cm (18 in.). The fixture has concentric black louvers with 0.6 cm (0.25 in.) spacing. Depending on the lamps used, the fixture provides up to 72 W (22 W UV-C) and has an irradiation zone of approximately 360°. Figure 4C shows a ceiling-mounted fixture designed for ceilings of 2.7 m (9 ft) or greater. It is made from steel and comes in either 45.7 cm (18 in.) or 91.4 cm (36 in.) lengths. It provides up to 72 W (22 W UV-C) nominal output and the lamp's average effective life is rated at 8,000 h (an average of approximately 20% UV output depreciation).

4.3 System Installation

Since 1950, several articles have described the number and location of UVGI fixtures needed in a room to provide an effective upper-room UVGI system. Buttolph and Haynes [1950] noted that fixtures should be distributed to provide as uniform upper-room irradiation as possible. For example, in small rooms, two 15 W units mounted on opposite walls are better than a single 30 W unit. Sylvania [1981] specified the number of germicidal lamps for upper room-air irradiation based on the area of the room. They recommended two 15 W lamps or one 30 W lamp (open-front or louvered fixtures) for a 9.3 m^2 (100 ft^2) room and up to thirteen 30 W lamps for a 204.8 m^2 (2,200 ft^2) floor area. These recommendations were based on a 90% upper-room deactivation of *E. coli*.

A general rule of thumb that has been used in recent years is that one 30 W fixture should be installed for every 18 m^2 (200 ft^2) of floor area or for every seven people in a room, whichever is greater [First et al. 1999a; Macher 1993; Riley 1988; Riley and Nardell 1989]. This rule provides the lamp power to be used; however, it fails to consider that the UVGI actually provided by a fixture varies considerably depending on the fixture(s) location in the room (e.g., wall-mounted versus ceiling-mounted), the type and number of louvers (if any), the reflectivity of building materials [Miller and Macher 2000; Miller et al. 2002; Xu 2001], the UV-C output of the lamps, and the emission characteristics of the fixtures.

Another rule of thumb currently used for installing upper-room UVGI systems involves the consideration of the UV-C output (W) produced by the UVGI lamps divided by the area of the room [Dunn 2003]. In this approach, in rooms with ceilings below 2.7 m (9 ft), where louvered fixtures must be used, the number of fixtures installed provides a UV-C irradiance of 1.1 W/m^2 (0.1 W/ft^2). This approach assumes that open fixtures provide approximately 2 times the UVGI irradiance as louvered fixtures [Dunn 2003; Miller et al.

2002]. Therefore, in rooms with 2.7 m (9 ft) ceilings or above, when fixtures without louvers can be used, fixtures are installed that provide 0.5 W/m^2 (0.05 W/ft^2). The fixtures that are installed using these guidelines are positioned to provide as uniform UVGI distribution in the upper air as practical.

The approach described above is based on the UV-C output of the UVGI lamps used by Riley et al. [1976]. Variations of this approach using the floor area or volume of a room may be useful in installing effective upper-room UVGI systems until more precise methods for installing these systems are developed. These suggested approaches are based on the information obtained in the study by Miller et al. [2002]. In this study, as noted previously, an average UV fluence rate of 42 µW/cm^2 in the upper room was obtained using five louvered UVGI fixtures (one in the center of the room and one in each corner). These fixtures provided for the inactivation of airborne *M. parafortuitum* at an average rate of (16 ± 1.2) eACH in decay experiments and reduced the room-average concentration of culturable airborne *M. parafortuitum* more than 90% in CG experiments at 50% RH. The five fixtures contained a total of 12 lamps that provided 216 W (66 W UV-C) nominal output. The test room was 2.5 m (8.1 ft) high with an area of 35.3 m^2 (380 ft^2), which corresponds to an average of 1.87 W/m^2 UV-C (0.17 W/ft^2 UV-C). An alternative approach is to install fixtures based on the volume of the room that is irradiated [Xu et al. 2005]. During the Miller et al. [2002] study described above, the depth of the UVGI band averaged approximately 30 cm (11.8 in.) with a total irradiated air volume of 10.4 m^3 (367.3 ft^3). This resulted in a power distribution of 6.3 W/m^3 UV-C (66 W UV-C per 10.4 cubic meters) or 0.18 W/ft^3 UV-C. Xu et al. [2005] suggest that to obtain the maximum benefit of an upper-room UVGI system, a reasonable volume of air in the upper zone should be irradiated with at least 6 W/m^3 UV-C.

Considering all parameters, the installation of UVGI fixtures in rooms with approximately 2.4 m (8 ft) ceilings that provide (1) 1.87 W/m^2 UV-C (0.17 W/ft^2 UV-C) irradiance in the floor space or (2) 6 W/m^3 UV-C (0.18 W/ft^3 UV-C) in the upper-UVGI zone may be effective in killing or inactivating airborne *M. tuberculosis*. For rooms with ceilings that are 2.7 m (9 ft.) or greater where fixtures without louvers can be used, there is a direct relationship between the size of the irradiated zone in a room and the amount of time droplet nuclei are exposed to UVGI. For example, if the size of the irradiated zone is doubled in a room, the length of time the average droplet nuclei is exposed to UVGI in the zone doubles [Beggs and Sleigh 2002]. Therefore, based on modeling, the irradiance can be reduced by 50% and the droplet nuclei will still receive the same dose.

As noted above, a number of factors must be evaluated when designing a safe and effective upper-room UVGI system. Consideration of all factors makes it difficult to determine the number and types of fixtures necessary. The room size and the amount of UVGI needed are important factors to consider when designing an upper-room UVGI system [Miller et al. 2002]. Other factors that need to be considered include the room geometry, volume, fixture output, and air mixing including room ventilation, and the placement of the supply diffusers and exhaust grills [First et al. 1999a; Xu 2001]. It is important to note that rooms with lower ceilings (e.g., 2.4 m [8 ft]) do not have the same volume of air available for irradiation as do rooms or areas with higher ceilings. Also, UVGI fixtures without louvers may be used in rooms or areas with higher ceilings and, as noted above, they generally provide higher UV irradiation levels than louvered fixtures. Therefore, rooms or areas with low ceilings that use louvered fixtures may need to use more lamps (or fixtures) to provide the same effectiveness as fixtures without louvers that are used in rooms or areas with relatively high ceilings. To provide as uniform a UVGI distribution in the upper air as possible, UVGI fixtures should be placed to reduce overlap between fixtures and maintain an even irradiance zone.

4.4 Installation and Lamp Maintenance

Problems that have been found by CDC/NIOSH in the installation and maintenance of upper-room UVGI systems are noted in Appendix A. They were considered in the development of these guidelines.

A professional who has received training on the installation and placement of UVGI lamp fixtures, and is knowledgeable about their operation should be consulted before procurement and installation of the system. The number of persons properly trained in the design of upper-room UVGI systems is limited. Persons who may be consulted include engineers, industrial hygienists, other health professionals [CDC 2005a], and radiation/health physics professionals.

Once the number and types of UVGI fixtures appropriate for the room or area have been determined, the fixtures need to be appropriately installed. Only qualified service technicians should install the systems.

1. Ideally, the UV lamps in a fixture should not be visible from any position in the room.

2. Fixtures should contain baffles or louvers appropriately positioned to direct UVGI to the upper room if the ceiling is relatively low (e.g., between 2.4 m [8 ft] and 2.7 m [9 ft]).

3. Electrical installation should be done in accordance with the National Electric Code, local regulations, and the manufacturer's instructions.

4. On/Off switches for the fixtures can be installed in the same location as room lighting switches. However, a safety switch should be used to ensure that a fixture(s) is not activated during maintenance.

5. Provisions should be made to ensure that fixtures are not turned off when they should be on. For example, simple keyed switches may be used.

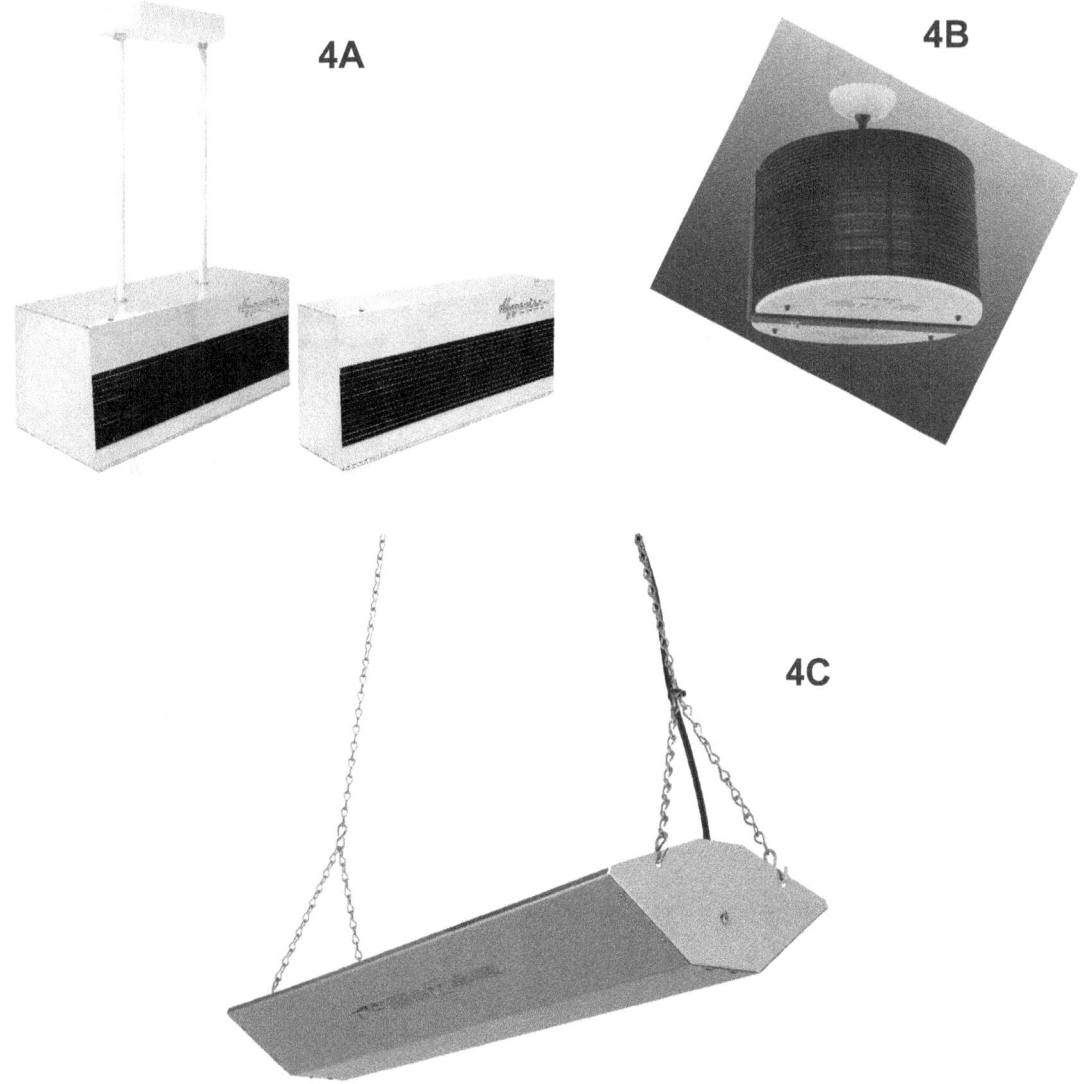

Figure 4. UVGI fixtures. **A**) Pendulum and wall mount fixtures, courtesy of Atlantic Ultraviolet Corporation, Hauppauge, NY. **B**) Pendulum and **C**) ceiling (pendant) fixtures, courtesy of Lumalier Lighting Design, Inc., Memphis TN.

6. If unshielded UVGI lamps are used in a room, entrance doors should have a warning sign, and a switching device should be installed to deactivate the lamps when the doors are opened. In some situations, the use of a motion sensor to deactivate the lamps when anyone enters the room may be appropriate.

7. In accordance with OSHA regulations (29 CFR 1910.145), warning signs should be placed on all UVGI fixtures and in situations where the UVGI exposure may be greater than the CDC/NIOSH REL. The warning signs should be written in the language of the affected persons.

> **CAUTION**
> **ULTRAVIOLET ENERGY: TURN OFF LAMPS**
> **BEFORE ENTERING UPPER ROOM**

General warning signs should be posted near the fixture(s) that contain the following information:

> **CAUTION**
> **ULTRAVIOLET ENERGY**
> **PROTECT EYES AND SKIN**

8. If UVGI measurements in the lower portion of the room exceed the CDC/NIOSH REL, all highly UV reflecting materials should be removed, replaced, or covered. As needed, UV-absorbing paints such as titanium dioxide can be used on ceilings and walls to minimize UVGI reflectance in occupied areas or the system can be redesigned.

4.5 UV Fixtures and Lamp Maintenance

All germicidal lamps experience a decrement in UV emission as they are used. Also, the output of the UV lamps may decrease due to accumulated dust. Therefore, lamps should be inspected periodically (e.g., quarterly) and cleaned if appropriate based on the operating environment or conditions. UVGI fixtures should be inspected and cleaned on the same basis as the UV lamps. All UVGI systems must be inactivated before workers enter the upper irradiated portion of a room or area. Employers should develop workplace procedures to ensure workers are not exposed to UVGI above the CDC/NIOSH REL. All personnel should receive training in exposure hazards. Maintenance personnel should receive additional safety training [CDC 2005a].

1. Turn off the upper-room UVGI system and let the lamps/fixtures cool.

2. Open the unit in accordance with the manufacturer's directions.

3. Remove the lamps from the fixture for cleaning. Handle the lamps only while wearing clean cotton gloves to prevent oil deposits from accumulating on the lamps and decreasing their emission efficiency.

4. Use a cloth dampened with alcohol (e.g., 91% isopropyl) to clean the lamps and reflectors—do not use water. Dry the lamps and reflectors with a soft cotton cloth to remove any residue while continuing to wear cotton gloves.

5. Lamps should be changed according to a fixed schedule based on the lamp manufacturers' recommendation. If feasible, group relamping should be done on a yearly basis. The lamp or ballast should also be replaced if the lamp stops glowing or flickers.

6. Close the fixture.

7. When all appropriate lamps have been replaced in the upper-room UVGI system, turn on (reenergize) the system and verify (e.g., visually) lamp operation and that (if present) all louvers are in the correct position. If necessary, UV-protective eyewear should be used when verifying the lamps are reenergized.

8. Document inspection, cleaning, and lamp replacement in a preventive maintenance logbook.

4.6 Personal Protective Equipment (PPE)

High levels of UVGI can occur close to the fixtures and the CDC/NIOSH REL can be exceeded in minutes or seconds (refer to Table 1). The appropriate PPE should be used (1) whenever UVGI measurements are being made in the irradiated portion(s) of a room or area (e.g., above 1.83 m [6 ft]), if exposures may occur in excess of the CDC/NIOSH REL or (2) if maintenance must be performed in the irradiated portion of a room where UVGI fixtures are activated. Workers should be trained in the use of PPE. PPE includes the following:

- UV-protective safety glasses with side shields or a face shield.

- Clothing made of tightly woven fabric.

- Covering opaque to UVGI such as barrier creams containing either zinc oxide or titanium dioxide for head, neck, face, and other exposed skin.

- Soft cotton gloves.

4.7 UVGI Measurements

In general, it has been recommended that the initial irradiance measurements of a newly installed UVGI system should not be made until the lamps have been burned in for at least 100 h due to rapid decline in emission power during this period [Dumyahn and First 1999; First et al. 1999a; Rea 2000; Shechmeister 1991]. However, the burn-in period may not be necessary and manufacturers' recommendations should be followed since some new model lamps do not undergo the initial decline in emission power. Logbooks should be maintained showing UVGI levels in the lower room containing information such as the date, time, and location of the measurements for each upper-room UVGI system, detector measurement readings, and adverse overexposure incidents. Irradiance measurements are generally required for two reasons: (1) validation that the system provides a sufficient UVGI level to ensure microbial inactivation and (2) determination of compliance with occupational safety and health guidelines. Currently, several methods are available for estimating UVGI in the upper room. These include the use of real-time measurements using broadband radiometers and spectroradiometers, chemical actinometry, and modeling as described in section 4.7.1. UV meters or dosimeters can be used for measuring worker exposure to UVGI in the lower (occupied) portion of the room.

Before an upper-room UVGI system is installed, the expected irradiance for all of the fixtures can be estimated using mathematical modeling or computational fluid dynamics (CFD) modeling as described in Appendix B. Dumyahn and First [1999] noted significant differences in UVGI emission patterns between individual fixtures of the same make and model. These differences were caused by variations in the (1) spacing and orientation of louvers (due to loose support rods), (2) mirror (reflector) shapes, and (3) orientation of lamp mounting brackets. Therefore, once the system has been installed, measurements should be made to provide an estimate of the actual average upper-room UVGI irradiance provided by the system. If the UVGI irradiance in the upper room is insufficient (e.g., less than 30 $\mu W/cm^2$ to 50 $\mu W/cm^2$), a number of options exist including reconfiguring existing fixtures or installation of additional fixtures. Detailed measurements only need to occur during the initial installation of the system or during refurbishment. Spot measurements can be done to verify lamp alignment or to check output on a quarterly basis.

Measurements should also be taken in the lower room to estimate the UVGI exposure level of room occupants. It is appropriate to make occupational UVGI exposure measurements (e.g., at workers' typical eye level) when the systems are initially installed or modified, lamps are replaced, or room surfaces are repainted or recovered. The data can then be compared with the CDC/NIOSH REL and, if needed, appropriate protective measures taken. Otherwise, occupational measurements may not be needed on a regular basis unless potential UVGI-related adverse health effects are observed.

4.7.1 Measurement Approaches

The primary instruments used to measure UVGI (flux density) are broadband radiometers and spectroradiometers. Generally, the UVGI systems used today are continuous output rather than pulsed sources. Measurements of continuous-wave UV sources can be made with photoelectric detectors (photomultiplier tubes and photodiodes) and thermal detectors such as thermopiles and pyroelectric detectors. These radiometers are used to measure radiant power and have calibrated output in irradiance. Often these types of meters have an integration function that permits radiant exposure to be measured over time. Detectors used for broadband instruments should display a wavelength-independent (flat) response in the spectral response region of interest and have a rapid response decrease outside the region of interest. Radiometers are flexible, smaller, and easier to use than spectroradiometers. However, their accuracy is dependent on ensuring that the filter used is appropriate to the diffuser. Also, the radiometer optical systems are easily contaminated by fingerprints and dirt. Regardless of the equipment used, it must be calibrated on a regular basis (e.g., yearly) by the manufacturer or other responsible party.

Spectroradiometers measure irradiance in a narrow wavelength region. They generally have wavelength-specific calibration factors to determine the actual spectral irradiance. Spectroradiometers generally are larger, more expensive, require special training, and take more time to set up and operate than radiometer systems. However, if used properly they provide a better estimate of the UVGI flux.

Before making measurements, it is important to understand how the meter functions and to ensure that the equipment is calibrated and operates in the appropriate wavelength region. The meter must be allowed to warm up for the recommended period of time, and the output indicator zeroed and checked for drift. In addition, if the evaluator is to measure upper-room UVGI levels, the evaluator must wear appropriate PPE since measurements may be taken near the activated UVGI fixtures.

4.8 Measuring UVGI Irradiance

4.8.1 Lower (Occupied Level)

As noted previously, measurements in the lower room should be taken to determine compliance with occupational safety and health guidelines. UVGI levels should be measured with a calibrated UV meter or dosimeter. Since eyes are the most sensitive organ to UVGI, measurements should be taken at eye level. The meter or dosimeter should be held at eye level and moved about the room.

The job functions, movement of workers, and time spent at the measurement locations in the room should be considered before measurements are taken and compared with occupational exposure limits. For example, if workers undertake most of their tasks sitting at a desk, the UVGI measurements may be taken at eye level from a sitting position. If most of a worker's day is spent standing, measurements should be taken at eye level for a standing person (e.g., 1.75 m above the floor). Several locations throughout the room or area should be selected to make the measurements. It should be noted that "hot spots" may occur where elevated irradiance levels will be found (e.g., caused by reflective surfaces, poor design, incorrect installation, or damaged units). A person trained in the use of a UV meter or dosimeter should make the readings.

Based on the UVGI levels obtained, compliance with occupational safety and health guidelines can be determined. Table 1 provides the CDC/NIOSH REL for UVGI (254 nm) exposure at selected time periods. The formula in Table 1 can be used to calculate different exposure times than those provided and exposure periods greater than 8 h. It should be noted that the REL is based on an average exposure to UVGI at an irradiance level and exposures throughout the workday will most likely vary [First et al. 2005]. A person knowledgeable in UVGI exposure readings should ascertain compliance with existing guidelines [ACGIH 2007; NIOSH 1972].

4.8.2 Upper-Room (Irradiated) Level

The present methods of measuring upper-room UVGI levels in the irradiated zone are cumbersome. However, to confirm the best possible configuration of UV lamps, distribution measurements should be made to ensure appropriate UVGI irradiance. One approach that has been used to estimate the UVGI irradiance in the upper room is to use a UVGI detector/radiometer system [Dumyahn and First 1999; Miller et al. 2002]. Since irradiance will vary at different locations in the room, several such measurements must be taken. Dumyahn and First [1999] evaluated the upper-room UVGI produced by individual fixtures using a radiometer at graduated distances and angles from the fixture's horizontal and vertical centerlines.

When more than one UVGI fixture is used in an upper-room UVGI system, estimating the total upper-air UVGI irradiance is more difficult. As noted above, airborne microorganisms move with rotational and translational trajectories in space. If a microorganism moves in a space that contains UVGI, then it is exposed from all directions. When using multiple UVGI fixtures in a room, the fluence is determined by adding the fluence attributed to each of the fixtures [Rudnick 2001]. Also, scattering and reflection of UVGI can increase complexity.

One method for estimating the additive effect of multiple fixtures at any given location has been suggested by Nardell et al. [2004]. Each fixture is turned on separately and the maximal irradiance measured at the point of interest with a photometer. The summed result of all of the individual fixtures should represent an approximation of the total irradiance at that point. An average upper-room irradiance level can be obtained if multiple readings are taken throughout the upper room.

A radiometer limits the measured radiation to that which is perpendicular to the detector surface. If droplet nuclei are considered to be located within a small cube, then the sum of the UVGI irradiance measurements made normal to each of the six sides of the cube can be viewed as an approximation of the total UV irradiance incident on the droplet nuclei at that location. This was the basis of a method used by Miller et al. [2002] to establish the upper-room UVGI levels when several UVGI fixtures were operating in a room. The floor of a room was divided into a number of square grids 0.6 m on a side. A radiometer was attached to a pole and used to document measurements. The pole was moved to the center of each grid, and the radiometer was placed in the center of the UV beam in the upper portion of the room (2.3 m above the floor)—one measurement was taken every 90 degrees by rotating the pole. During this process, the pole was horizontally and vertically aligned.

Measurements were also taken at the center of each grid in both the up and down directions. All measurements were made at least 1.2 m away from each fixture in all applicable directions to ensure the UVGI levels were not too high for the detector to function properly. The sums of the measurements were a reasonable estimation of the UVGI average obtained using actinometry [Rahn et al. 1999].

Although not every square will have the same level, there should be a relatively uniform spatial distribution. Ideally, using this method, the average of all of the radiometer measurements should be 30 $\mu W/cm^2$ to 50 $\mu W/cm^2$ (see section 3).

5 RESEARCH NEEDS

Considerable progress has been made in understanding the parameters that affect upper-room UVGI systems. Guidelines for the installation of effective upper-room UVGI systems are based mainly on recent experimental laboratory studies. However, these studies are limited because no two situations where upper-room systems are to be installed will be exactly like the conditions evaluated in the experimental studies. For example, parameters (e.g., ventilation, air mixing, humidity) will differ considerably from those examined in these guidelines. Also, room characteristics such as configuration, number of occupants and their activities, and placement of furniture have not been fully considered in the guidelines. Since much is yet to be determined, it is expected that additional research in the areas noted below will significantly add to the knowledge necessary to design effective and safe upper-room UVGI systems.

5.1 UVGI Measurements

5.1.1 Upper-Room Measurements

A number of methods can be used to determine the average upper-room UVGI irradiance. However, these methods tend to be cumbersome and are best used in research settings. A method that will quickly and easily provide this information is needed. An optical firm (GigaHertz Optik) has developed a cylindrical UV probe for taking 360° measurements. Additional research needs to be performed and published that evaluates the accuracy of the probe.

5.1.2 Lower-Room Measurements

Methods for estimating UV irradiance levels for workers who frequently move about during the course of their work need to be developed and validated. For example, personal dosimeters that reflect workers' exposure at eye level would be useful in determining compliance with occupational safety and public health guidelines.

5.2 Room Air Mixing

Methods need to be developed that determine if existing room air mixing is sufficient for UVGI effectiveness. Also, research should be conducted to determine if the use of mixing fans negatively impacts on the intended design of the mechanical ventilation systems or has a negative impact on infection control measures.

5.3 Relative Humidity

There is some indication that low RH (below 25%) may adversely affect UVGI systems and their ability to kill or inactivate airborne bacteria. Additional research is needed in this area. Also, research in full-scale rooms to better ascertain the effects of high humidity (e.g., 80% RH) on airborne microorganisms is needed.

5.4 Microbial Sensitivity

This document is concerned with the effect of upper-room UVGI systems on *M. tuberculosis*. Presently, tests to determine the relative sensitivity of microorganisms to UVGI are not standardized between different laboratories. Laboratory testing guidelines should be developed to ensure that they are reproducible and reflect real world situations.

5.5 Safety and Health Guidelines

The protection provided by sunscreen against UV-C radiation has not been determined. Research is needed to determine if sunscreen (e.g., SPF 15 or higher) will protect workers exposed to germicidal irradiation above the CDC/NIOSH REL.

5.6 Testing and Commissioning

Laboratory testing on the efficacy of UVGI upper-room systems should be standardized. Protocols for testing and validating upper-room UVGI systems need to be developed to ensure that the systems perform as designed [Kowalski and Bahnfleth 2004]. A system that provides training certification for system designers should be developed.

5.7 Performance Guidelines

Performance guidelines need to be developed for certifying UV fixtures to ensure their safe use. In addition, appropriate guidelines are needed for UV meters such as wavelength resolution and accuracy, limits of error, allowable drift, stray light rejection, and sensitivity for broadband and narrowband radiometers and spectroradiometers.

5.8 Mechanical Ventilation

Experimental research has indicated that mechanical ventilation up to 6 ACH does not have a significant effect on the effectiveness of upper-room UVGI systems. Does mechanical ventilation greater than 6 ACH decrease the effectiveness of upper-room UVGI systems?

5.9 Planning Guidelines

Guidelines need to be developed on the most practical method for planning effective UVGI systems in a variety of rooms or areas. As described in Appendix B, mathematical models have been used to provide estimates of the UVGI irradiance in the upper air. CFD modeling can theoretically be used to consider many of the variables associated with installing an upper-room UVGI system and provide an estimate of the UVGI dose received by droplet nuclei. To increase the accuracy of mathematical models, additional research is needed on the assumptions that they are based on.

5.10 Photoreactivation

Experiments involving PR of microorganisms in full-scale test rooms should be conducted.

5.11 Temperature

Studies on the effectiveness of UVGI on airborne bacteria over a wide range of temperatures should be conducted.

5.12 UVGI Effectiveness

In "real world" situations, potentially infectious droplet nuclei will vary in size and may be coated with sputum. Both of these factors may decrease UVGI effectiveness. Some laboratory research has been done to evaluate these parameters [Ko et al. 2000; Lai et al. 2004; First et al. 2007]. However, additional research needs to be done to further characterize microbial susceptibility to UVGI based on (a) the size of respirable (up to 5 µm) droplet nuclei and (b) droplet nuclei coated with actual or simulated sputum.

5.13 UV Fixtures

New UV fixtures designed to provide increased UVGI in the upper room while limiting the UVGI exposure in the lower (occupied) room need to be developed [First et al. 2007]. Also, a standard test protocol to measure the UV power output of commercially available UV fixtures is needed [Rudnick and First 2007].

REFERENCES

ACCP (American College of Chest Physicians) [1995]. Institutional control measures for tuberculosis in the era of multiple drug resistance: ACCP/ATS consensus conference. Chest *108*(6):1690–1710.

ACGIH [2004]. Industrial ventilation: a manual of recommended practice. 25th ed. Cincinnati, OH: American Conference of Governmental Industrial Hygienists.

ACGIH [2007]. Ultraviolet radiation. In: 2007 TLVs and BEIs: based on the documentation of the threshold limit values for chemical substances and physical agents & biological exposure indices. Cincinnati, OH: American Conference of Governmental Industrial Hygienists.

AIA (American Institute of Architects) [2006]. Guidelines for design and construction of health care facilities. Washington, DC: American Institute of Architects.

ASHRAE [2004]. ANSI/ASHRAE standard 55–2004: thermal environmental conditions for human occupancy. Atlanta, GA: American Society of Heating, Refrigerating, and Air-Conditioning Engineers.

ASHRAE [2005]. Thermal comfort. In: 2005 ASHRAE handbook—fundamentals. Atlanta, GA: American Society of Heating, Refrigerating, and Air-Conditioning Engineers.

ASHRAE [2007]. Health care facilities. In: 2007 ASHRAE handbook—HVAC applications. Atlanta, GA: American Society of Heating, Refrigerating, and Air-Conditioning Engineers.

Baker MN [1981]. The quest for pure water: the history of water purification from the earliest records to the twentieth century. 2nd ed. Vol. 1. American Water Works Association, pp. 354–356.

Beggs CB, Sleigh PA [2002]. A quantitative method for evaluating the germicidal effect of upper room UV fields. J Aerosol Sci *33*(12):1681–1699.

Benenson AS, ed. [1990]. Control of communicable diseases in man. 15th ed. Washington, DC: American Public Health Association.

Boudreau Y, Decker JA, Burton N [1995]. Hazard evaluation and technical assistance report: Jackson Memorial Hospital, Miami, FL. Cincinnati, OH: U.S. Department of Health and Human Services, Centers for Disease Control and Prevention, National Institute for Occupational Safety and Health, NIOSH HETA Report No. 91–0187–2544, NTIS No. PB96–209762. [www.cdc.gov/niosh/hhe/reports/pdfs/1991-0187-2544.pdf]

Brickner PW, Vincent RL, First M, Nardell E, Murray M, Kaufman W [2003]. The application of ultraviolet germicidal irradiation to control transmission of airborne disease:

bioterrorism countermeasure. Public Health Rep *118*(2):99–114. [www.pubmedcentral.nih.gov/picrender.fcgi?artid=1497517&blobtype=pdf]

Brubacher J, Hoffman RS [1996]. Hazards of ultraviolet lighting used for tuberculosis control. Chest *109*(2):582–583.

Buchta TM, Klein MK, Martinez K [1993]. Hazard evaluation and technical assistance report: 44th Street Independence Support Center, New York, NY. Cincinnati, OH: U.S. Department of Health and Human Services, Centers for Disease Control and Prevention, National Institute for Occupational Safety and Health, NIOSH HETA Report No. 92–0320–2357, NTIS No. PB94–152154. [www.cdc.gov/niosh/hhe/reports/pdfs/1992-0320-2357.pdf]

Burton NC [1995]. Closeout letter of August 3 to County of Orange Department of Health, Goshen, NY. Cincinnati, OH: U.S. Department of Health and Human Services, Centers for Disease Control and Prevention, National Institute for Occupational Safety and Health, Division of Surveillance, Hazard Evaluations and Field Studies, HETA No. 93–0746. Unpublished.

Burton NC, Martinez KF [1997]. Closeout letter of November 10 to Department of Veterans Affairs Medical Center, New York, NY. Cincinnati, OH: U.S. Department of Health and Human Services, Centers for Disease Control and Prevention, National Institute for Occupational Safety and Health, Division of Surveillance, Hazard Evaluations and Field Studies, HETA No. 93–0652. Unpublished.

Burwen DR, Bloch AB, Griffin LD, Ciesielski CA, Stern HA, Onorato IM [1995]. National trends in the concurrence of tuberculosis and acquired immunodeficiency syndrome. Arch Intern Med *155*(12):1281–1286.

Buttolph LJ, Haynes H [1950]. Ultraviolet air sanitation. Cleveland, OH: General Electric, Bulletin LD–11.

Cantwell MF, Snider DE Jr., Cauthen GM, Onorato IM [1994]. Epidemiology of tuberculosis in the United States, 1985 through 1992. JAMA *272*(7):535–539.

Casarett AP [1968]. Radiation biology. Englewood Cliffs, NJ: Prentice-Hall.

CDC (Centers for Disease Control and Prevention) [1994]. Guidelines for preventing the transmission of *Mycobacterium tuberculosis* in health-care facilities. MMWR *43*(RR–13). [www.cdc.gov/mmwr/PDF/rr/rr4313.pdf]

CDC (Centers for Disease Control and Prevention) [2000]. Targeted tuberculin testing and treatment of latent tuberculosis infection. MMWR *49*(RR–6):1–54. [www.cdc.gov/mmwr/PDF/rr/rr4906.pdf]

CDC (Centers for Disease Control and Prevention) [2002]. Tuberculosis morbidity among U.S.-born and foreign-born populations—United States, 2000. MMWR *51*(5):101–104. [www.cdc.gov/mmwr/preview/mmwrhtml/mm5105a3.htm]

CDC (Centers for Disease Control and Prevention) [2003]. Trends in tuberculosis morbidity—United States, 1992–2002. MMWR *52*(11):217–222. [www.cdc.gov/mmwr/preview/mmwrhtml/mm5211a2.htm]

CDC (Centers for Disease Control and Prevention) [2004]. Trends in tuberculosis—United States, 1998–2003. MMWR *53*(10):209–214. [www.cdc.gov/mmwr/preview/mmwrhtml/mm5310a2.htm]

CDC (Centers for Disease Control and Prevention) [2005a]. Guidelines for preventing the transmission of *Mycobacterium tuberculosis* in health-care settings, 2005. MMWR *54*(RR–17). [www.cdc.gov/mmwr/PDF/rr/rr5417.pdf]

CDC (Centers for Disease Control and Prevention) [2005b]. Trends in tuberculosis—United States, 2004. MMWR *54*(10):245–249. [www.cdc.gov/mmwr/preview/mmwrhtml/mm5410a2.htm]

CFR. Code of Federal Regulations. Washington, DC: U.S. Government Printing Office, Office of the Federal Register.

CIE (International Commission on Illumination) [1987]. International lighting vocabulary. 4th ed. Geneva: Bureau Central de la Commission Electrotechnique Internationale, CIE Publication 17.4.

Collins FM [1971]. Relative susceptibility of acid-fast and non-acid-fast bacteria to ultraviolet light. Appl Microbiol *21*(3):411–413. [www.pubmedcentral.nih.gov/picrender.fcgi?artid=377194&blobtype=pdf]

David HL, Jones WD Jr., Newman CM [1971]. Ultraviolet light inactivation and photoreactivation in the mycobacteria. Infect Immun *4*(3):318–319. [www.pubmedcentral.nih.gov/picrender.fcgi?artid=416306&blobtype=pdf]

Decker JA [1993]. Hazard evaluation and technical assistance report: Manatee Memorial Hospital, Bradenton, FL. Cincinnati, OH: U.S. Department of Health and Human Services, Centers for Disease Control and Prevention, National Institute for Occupational Safety and Health, NIOSH HETA Report No. 93–282–2303, NTIS No. PB93–234441. [www.cdc.gov/niosh/hhe/reports/pdfs/1993-0282-2303.pdf]

Diffey BL [1991]. Solar ultraviolet radiation effects on biological systems. Phys Med Biol *36*(3):299–328.

Downes A, Blunt TP [1877]. Researches on the effect of light upon bacteria and other organisms. Proc R Soc Lond 26:488–500.

Duguid JP [1946]. The size and the duration of air-carriage of respiratory droplets and droplet-nuclei. J Hyg *44*:471–479.

Dumyahn T, First M [1999]. Characterization of ultraviolet upper room air disinfection devices. Am Ind Hyg Assoc J *60*(2):219–227.

Dunn C Sr. [2003]. Telephone conversation on December 4 between C. Dunn, Sr., Lumalier Inc., Memphis, TN, and J. Whalen, Division of Federal Occupational Health, under contract with Division of Applied Research and Technology, National Institute for Occupational Safety and Health, Centers for Disease Control and Prevention, U.S. Department of Health and Human Services.

62 Fed. Reg. 54159 [1997]. Occupational Safety and Health Administration: occupational exposure to tuberculosis; proposed rule. Washington, DC: U.S. Government Printing Office, Office of the Federal Register. [www.osha.gov/pls/oshaweb/owadisp.show_document?p_table=FEDERAL_REGISTER&p_id=13717]

Fennelly KP, Martyny JW, Fulton KE, Orme IM, Cave DM, Heifets LB [2004]. Cough-generated aerosols of *Mycobacterium tuberculosis*: a new method to study infectiousness. Am J Respir Crit Care Med *169*(5):604–609.

Field MJ, ed. [2001]. Tuberculosis in the workplace. Washington, DC: National Academy Press.

First MW, Nardell EA, Chaisson W, Riley R [1999a]. Guidelines for the application of upper-room ultraviolet germicidal irradiation for preventing transmission of airborne contagion—part I: basic principles. ASHRAE Trans *105*:869–876.

First MW, Nardell EA, Chaisson W, Riley R [1999b]. Guidelines for the application of upper-room ultraviolet germicidal irradiation for preventing transmission of airborne contagion—part II: design and operation guidance. ASHRAE Trans *105*:877–887.

First MW, Weker RA, Yasui S, Nardell EA [2005]. Monitoring human exposures to upper-room germicidal ultraviolet irradiation. J Occup Environ Hyg *2*(5):285–292.

First M, Rudnick SN, Banahan KF, Vincent RL, Brickner PW [2007]. Fundamental factors affecting upper-room ultraviolet germicidal irradiation—part I: experimental. J Occup Environ Hyg *4*(5):321–331.

Gates FL [1929a]. A study of the bactericidal action of ultra violet light—I: the reaction to monochromatic radiations. J Gen Physiol *13*(2):231–248.

Gates FL [1929b]. A study of the bactericidal action of ultra violet light—II: the effect of various environmental factors and conditions. J Gen Physiol *13*(2):249–260.

GE (General Electric) [2005]. Germicidal lamp technical sheets. [www.gelighting.com/na/business_lighting/education_resources/literature_library/product_brochures/specialty/downloads/germicidal/germicidal_tech_sheets.pdf]. Date accessed: February 2007.

Hollaender A [1943]. Effect of long ultraviolet and short visible radiation (3500 to 4900Å) on *Escherichia coli*. J Bacteriol *46*(6):531–541. [www.pubmedcentral.nih.gov/picrender.fcgi?artid=373856&blobtype=pdf]

IAEA [1978]. Particle size analysis in estimating the significance of airborne contamination. Vienna: International Atomic Energy Agency. Technical Reports Series No. 179.

Jensen MM [1964]. Inactivation of airborne viruses by ultraviolet irradiation. Appl Microbiol *12*(5):418–420. [www.pubmedcentral.nih.gov/picrender.fcgi?artid=1058147&blobtype=pdf]

Jensen PA, Hayden CS II, Burroughs GE, Hughes RT [1998]. Assessment of the health hazard associated with the use of smoke tubes in healthcare facilities. Appl Occup Environ Hyg *13*(3):172–176.

Kethley TW [1973]. Feasibility study of germicidal UV lamps for air disinfection in simulated patient care rooms. Unpublished paper presented at the American Public Health Association Conference, Section on Environment, San Francisco, CA, November 7.

Kethley TW, Branch K [1972]. Ultraviolet lamps for room air disinfection: effect of sampling location and particle size of bacterial aerosol. Arch Environ Health *25*(3):205–214.

Ko G, First MW, Burge HA [2000]. Influence of relative humidity on particle size and UV sensitivity of *Serratia marcescens* and *Mycobacterium bovis* BCG aerosols. Tuber Lung Dis *80*(4/5):217–228.

Ko G, First MW, Burge HA [2002]. The characterization of upper-room ultraviolet germicidal irradiation in inactivating airborne microorganisms. Environ Health Perspect *110*(1):95–101. [www.pubmedcentral.nih.gov/picrender.fcgi?artid=1240698&blobtype=pdf]

Kowalski WJ [2006]. Aerobiological engineering handbook: a guide to airborne disease control technologies. New York: McGraw-Hill.

Kowalski WJ, Bahnfleth WP [2000]. UVGI design basics for air and surface disinfection. HPAC Eng *72*(1):100–110.

Kowalski WJ, Bahnfleth WP [2004]. Proposed standards and guidelines for UVGI air disinfection. IUVA News *6*(1):20–25.

Kowalski WJ, Bahnfleth WP, Witham DL, Severin BF, Whittam TS [2000]. Mathematical modeling of ultraviolet germicidal irradiation for air disinfection. Quant Microbiol *2*(3):249–270.

Kujundzic E, Matalkah F, Howard CJ, Hernandez M, Miller SL [2006]. UV air cleaners and upper-room air ultraviolet germicidal irradiation for controlling airborne bacteria and fungal spores. J Occup Environ Hyg *3*(10):536–546.

Lai KM, Burge HA, First MW [2004]. Size and UV germicidal irradiation susceptibility of *Serratia marcescens* when aerosolized from different suspending media. Appl Environ Microbiol *70*(4):2021–2027. [http://www.pubmedcentral.nih.gov/picrender.fcgi?artid=383042&blobtype=pdf]

Lidwell OM [1994]. Ultraviolet radiation and the control of airborne contamination in the operating room. J Hosp Infect *28*(4):245–248.

Loudon RG, Bumgarner LR, Lacy J, Coffman GK [1969]. Aerial transmission of mycobacteria. Am Rev Respir Dis *100*(2):165–171.

Luckiesh M [1946]. Applications of germicidal, erythemal and infrared energy. New York: D. Van Nostrand.

Luckiesh M, Holladay LL [1942]. Designing installations of germicidal lamps for occupied rooms. Gen Electric Rev *45*(6):343–349.

Macher JM [1993]. The use of germicidal lamps to control tuberculosis in healthcare facilities. Infect Control Hosp Epidemiol *14*(12):723–729.

Manangan LP, Collazo ER, Tokars J, Paul S, Jarvis WR [1999]. Trends in compliance with the guidelines for preventing the transmission of *Mycobacterium tuberculosis* among New Jersey hospitals, 1989 to 1996. Infect Control Hosp Epidemiol *20*(5):337–340.

Martinez KF [1995a]. Letter of May 19 to National Jewish Center for Immunology and Respiratory Medicine, Denver, CO. Cincinnati, OH: U.S. Department of Health and Human Services, Centers for Disease Control and Prevention, National Institute for Occupational Safety and Health, Division of Surveillance, Hazard Evaluations and Field Studies, HETA No. 92–0161. Unpublished.

Martinez KF [1995b]. Letter of September 26 to Tri-County Health Department, Commerce City, CO. Cincinnati, OH: U.S. Department of Health and Human Services, Centers for Disease Control and Prevention, National Institute for Occupational Safety and Health, Division of Surveillance, Hazard Evaluations and Field Studies, HETA No. 92–0161. Unpublished.

Martinez KF [1999]. Closeout letter of March 23 to Instituto Nacional de Salud Publica, Cuernavaca, Morelos, Mexico. Cincinnati, OH: U.S. Department of Health and Human Services, Centers for Disease Control and Prevention, National Institute for Occupational Safety and Health, Division of Surveillance, Hazard Evaluations and Field Studies, HETA No. 98–0296. Unpublished.

McLean RL [1961]. General discussion: the mechanism of spread of Asian influenza. Am Rev Respir Dis *83*:36–38.

Memarzadeh F [2000]. Assessing the efficacy of ultraviolet germicidal irradiation and ventilation in removing *Mycobacterium tuberculosis*. Bethesda, MD: U.S. Department of Health and Human Services, National Institutes of Health, Office of Research Services, Division of Engineering Services. [http://orf.od.nih.gov/PoliciesAndGuidelines/Bioenvironmental/assessubg_cover.htm]

Menzies D, Fanning A, Yuan L, Fitzgerald M [1995]. Tuberculosis among health care workers. N Engl J Med *332*(2):92–98.

Miller RV, Jeffrey W, Mitchell D, Elasri M [1999]. Bacterial responses to ultraviolet light. ASM News *65*(8):535–541.

Miller SL, Macher JM [2000]. Evaluation of a methodology for quantifying the effect of room air ultraviolet germicidal irradiation on airborne bacteria. Aerosol Sci Technol *33*(3):274–295.

Miller SL, Hernandez M, Fennelly K, Martyny J, Macher J, Kujundzic E, Xu P, Fabian P, Peccia J, Howard C [2002]. Efficacy of ultraviolet irradiation in controlling the spread of tuberculosis. Cincinnati, OH: U.S. Department of Health and Human Services, Centers for Disease Control and Prevention, National Institute for Occupational Safety and Health, final report, contract no. 200–97–2602; NTIS No. PB2003–103816. [www.cdc.gov/niosh/reports/contract/pdfs/ultrairrTB.pdf]

Moss CE, Seitz T [1990]. Hazard evaluation and technical assistance report: San Francisco General Hospital and Medical Center, San Francisco, CA. Cincinnati, OH: U.S. Department of Health and Human Services, Centers for Disease Control, National Institute for Occupational Safety and Health, NIOSH HETA Report No. 90–122–L2073, NTIS No. PB91–169359. [www.cdc.gov/niosh/hhe/reports/pdfs/1990-0122-L2073.pdf]

Moss CE, Seitz T [1991]. Hazard evaluation and technical assistance report: John C. Murphy Family Health Center, Berkeley, MO. Cincinnati, OH: U.S. Department of Health and Human Services, Centers for Disease Control, National Institute for Occupational Safety and Health, NIOSH HETA Report No. 91–148–2236, NTIS No. PB93–119956. [www.cdc.gov/niosh/hhe/reports/pdfs/1991-0148-2236.pdf]

Murray WE [1987]. Closeout letter of April 21 to New York State Department of Health, Albany, NY. Cincinnati, OH: U.S. Department of Health and Human Services, Centers for Disease Control, National Institute for Occupational Safety and Health, Division of Surveillance, Hazard Evaluations and Field Studies, HETA No. 86–169. Unpublished.

Mutchler JE [1973]. Principles of ventilation. In: The industrial environment—its evaluation and control. Cincinnati, OH: U.S. Department of Health, Education, and Welfare, Public Health Service, Center for Disease Control, National Institute for Occupational Safety and Health, DHEW (NIOSH) Publication No. 74–117.

Nardell EA [1995]. Interrupting transmission from patients with unsuspected tuberculosis: a unique role for upper-room ultraviolet air disinfection. Am J Infect Control *23*(2):156–164.

Nardell EA, Riley RL [1992]. A new ultraviolet germicidal irradiation (UVGI) fixture design for upper room air disinfection with low ceilings. Abstract. In: Program and abstracts of the World Congress on Tuberculosis, Bethesda, MD, November 16–19, p. 38.

Nardell EA, Keegan J, Cheney SA, Etkind SC [1991]. Airborne infection: theoretical limits of protection achievable by building ventilation. Am Rev Respir Dis *144*(2):302–306.

Nardell E (enardell@pih.org), First M, Rudnick S [2004]. Review of document "Engineering controls for tuberculosis: upper-air germicidal irradiation." Private e-mail message to Scott Earnest (GSE0@cdc.gov), December 17.

Nardell EA, Bucher SJ, Brickner PW, Wang C, Vincent RL, Becan-McBride K, James MA, Michael M, Wright JA [2008]. Safety of upper-room ultraviolet germicidal air disinfection for room occupants: results from the tuberculosis ultraviolet shelter study. Public Health Rep *123*(1):52–60.

Nicas M, Miller SL [1999]. A multi-zone model evaluation of the efficacy of upper-room air ultraviolet germicidal irradiation. Appl Occup Environ Hyg *14*(5):317–328.

NIOSH [1972]. Criteria for a recommended standard: occupational exposure to ultraviolet radiation. Rockville, MD: U.S. Department of Health, Education, and Welfare, Public Health Service, Health Services and Mental Health Administration, National Institute for Occupational Safety and Health, DHEW (NIOSH) Publication No. HSM 73–11009, NTIS No. PB–214268. [www.cdc.gov/niosh/73-11009.html]

Noakes C, Beggs C, Sleigh A [2004]. Evaluating upper room UVGI systems. ASHRAE IAQ Appl *2004*(Fall):17–20.

Osram [2005]. UV lamps. [http://catalog.myosram.com]. Date accessed: February 2007.

Papineni RS, Rosenthal FS [1997]. The size distribution of droplets in the exhaled breath of healthy human subjects. J Aerosol Med *10*(2):105–116.

Peccia JL [2000]. The response of airborne bacteria to ultraviolet germicidal radiation. [Thesis (Ph.D.)]. Boulder, CO: University of Colorado, Department of Civil, Environmental, and Architectural Engineering.

Peccia J, Hernandez M [2001]. Photoreactivation in airborne *Mycobacterium parafortuitum*. Appl Environ Microbiol *67*(9):4225–4232. [www.pubmedcentral.nih.gov/picrender.fcgi?artid=93151&blobtype=pdf]

Peccia J, Werth HM, Miller S, Hernandez M [2001]. Effects of relative humidity on the ultraviolet induced inactivation of airborne bacteria. Aerosol Sci Technol *35*(3):728–740.

Perkins JE, Bahlke AM, Silverman HF [1947]. Effect of ultra-violet irradiation of classrooms on spread of measles in large rural central schools: preliminary report. Am J Public Health Nations Health *37*(5):529–537. [www.pubmedcentral.nih.gov/picrender.fcgi?artid=1623610&blobtype=pdf]

Persily AK [1988]. Tracer gas techniques for studying building air exchange. Gaithersburg, MD: U.S. Department of Commerce, National Bureau of Standards, NBS Report No. NBSIR 88–3708.

Philips [1992]. Disinfection by UV-radiation. Philips Lighting Division, Booklet 3222 C34 00671.

Philips [2006]. Ultraviolet purification application information. The Netherlands: Philips Lighting B.V. [www.lighting.philips.com/gl_en/global_sites/application/water_purification/pdfs/uvp_application_brochure.pdf]

Rahn RO [1997]. Potassium iodide as a chemical actinometer for 254 nm radiation: use of iodate as an electron scavenger. Photochem Photobiol *66*(4):450–455.

Rahn RO, Xu P, Miller SL [1999]. Dosimetry of room-air germicidal (254 nm) radiation using spherical actinometry. Photochem Photobiol *70*(3):314–318.

Rea MS, ed. [2000]. The IESNA lighting handbook: reference & application. 9th ed. New York: Illuminating Engineering Society of North America.

Rentschler HC, Nagy R [1942]. Bactericidal action of ultraviolet radiation on air-borne organisms. J Bacteriol *44*(1):85–94. [www.pubmedcentral.nih.gov/picrender.fcgi?artid=373652&blobtype=pdf]

Rentschler HC, Nagy R, Mouromseff G [1941]. Bactericidal effect of ultraviolet radiation. J Bacteriol *41*(6):745–774. [www.pubmedcentral.nih.gov/picrender.fcgi?artid=374733&blobtype=pdf]

Riley RL [1988]. Ultraviolet air disinfection for control of respiratory contagion. In: Kundsin RB, ed. Architectural design and indoor microbial pollution. New York: Oxford University Press, pp. 174–197.

Riley RL, Kaufman JE [1971]. Air disinfection in corridors by upper air irradiation with ultraviolet. Arch Environ Health *22*(5):551–553.

Riley RL, Kaufman JE [1972]. Effect of relative humidity on the inactivation of airborne *Serratia marcescens* by ultraviolet radiation. Appl Microbiol *23*(6):1113–1120. [www.pubmedcentral.nih.gov/picrender.fcgi?artid=380516&blobtype=pdf]

Riley RL, Nardell EA [1989]. Clearing the air: the theory and application of ultraviolet air disinfection. Am Rev Respir Dis *139*(5):1286–1294.

Riley RL, O'Grady F [1961]. Airborne infection: transmission and control. New York: Macmillan, pp. 134–140.

Riley RL, Permutt S [1971]. Room air disinfection by ultraviolet irradiation of upper air: air mixing and germicidal effectiveness. Arch Environ Health *22*(2):208–219.

Riley RL, Wells WF, Mills CC, Nyka W, McLean RL [1957]. Air hygiene in tuberculosis: quantitative studies of infectivity and control in a pilot ward. Am Rev Tuberc *75*(3):420–431.

Riley RL, Mills CC, Nyka W, Weinstock N, Storey PB, Sultan LU, Riley MC, Wells WF [1959]. Aerial dissemination of pulmonary tuberculosis: a two-year study of contagion in a tuberculosis ward. Am J Hyg *70*:185–196.

Riley RL, Mills CC, O'Grady F, Sultan LU, Wittstadt F, Shivpuri DN [1962]. Infectiousness of air from a tuberculosis ward—ultraviolet irradiation of infected air: comparative infectiousness of different patients. Am Rev Respir Dis 85(4):511–525.

Riley RL, Permutt S, Kaufman JE [1971]. Convection, air mixing, and ultraviolet air disinfection in rooms. Arch Environ Health 22(2):200–207.

Riley RL, Knight M, Middlebrook G [1976]. Ultraviolet susceptibility of BCG and virulent tubercle bacilli. Am Rev Respir Dis 113(4):413–418.

Rudnick SN [2001]. Predicting the ultraviolet radiation distribution in a room with multi-louvered germicidal fixtures. AIHAJ 62(4):434–445.

Rudnick SN, First MW [2007]. Fundamental factors affecting upper-room ultraviolet germicidal irradiation—part II: predicting effectiveness. J Occup Environ Hyg 4(5):352–362.

Schwarz T [1998]. UV light affects cell membrane and cytoplasmic targets. J Photochem Photobiol B 44(2):91–96.

Segal-Maurer S, Kalkut GE [1994]. Environmental control of tuberculosis: continuing controversy. Clin Infect Dis 19(2):299–308.

Seitz T [1992]. Hazard evaluation and technical assistance report: Onondaga County Medical Examiner's Office, Syracuse, NY. Cincinnati, OH: U.S. Department of Health and Human Services, Centers for Disease Control, National Institute for Occupational Safety and Health, NIOSH HETA Report No. 92–171–2255, NTIS No. PB93–136356. [www.cdc.gov/niosh/hhe/reports/pdfs/1992-0171-2255.pdf]

Seitz TA, Boudreau Y, Martinez KF [2000]. Hazard evaluation and technical assistance report: Hawaii State Department of Health, Honolulu, HI. Cincinnati, OH: U.S. Department of Health and Human Services, Centers for Disease Control and Prevention, National Institute for Occupational Safety and Health, NIOSH HETA Report No. 2000–0040–2800, NTIS No. PB2005–104929. [www.cdc.gov/niosh/hhe/reports/pdfs/2000-0040-2800.pdf]

Setlow JK [1966]. The molecular basis of biological effects of ultraviolet radiation and photoreactivation. Curr Top Radiat Res 2:195–248.

Setlow RB [1997]. DNA damage and repair: a photobiological odyssey. Photochem Photobiol 65S:119S–122S.

Setlow RB, Setlow JK [1962]. Evidence that ultraviolet-induced thymine dimers in DNA cause biological damage. Proc Natl Acad Sci U S A 48(7):1250–1257. [www.pubmedcentral.nih.gov/picrender.fcgi?artid=220940&blobtype=pdf]

Sharp DG [1939]. The lethal action of short ultraviolet rays on several common pathogenic bacteria. J Bacteriol 37(4):447–460. [www.pubmedcentral.nih.gov/picrender.fcgi?artid=374478&blobtype=pdf]

Shechmeister IL [1991]. Sterilization by ultraviolet radiation. In: Block SS, ed. Disinfection, sterilization, and preservation. 4th ed. Philadelphia: Lea & Febiger, pp. 553–565.

Stead WW, Yeung C, Hartnett C [1996]. Probable role of ultraviolet irradiation in preventing transmission of tuberculosis: a case study. Infect Control Hosp Epidemiol *17*(1):11–13.

Sylvania [1981]. Germicidal and short-wave radiation. GTE Products Corp., Sylvania Engineering Bulletin 0–344.

Talbot EA, Jensen P, Moffat HJ, Wells CD [2002]. Occupational risk from ultraviolet germicidal irradiation (UVGI) lamps. Int J Tuberc Lung Dis *6*(8):738–741.

Tubbs RL, Bresler FT [1992]. Hazard evaluation and technical assistance report: A.G. Holley State Hospital, Lantana, FL. Cincinnati, OH: U.S. Department of Health and Human Services, Centers for Disease Control and Prevention, National Institute for Occupational Safety and Health, NIOSH HETA Report No. 92–215–2268, NTIS No. PB93–214021. [www.cdc.gov/niosh/hhe/reports/pdfs/1992-0215-2268.pdf]

VanOsdell D, Foarde K [2002]. Defining the effectiveness of UV lamps installed in circulating air ductwork. Arlington, VA: Air-Conditioning and Refrigeration Technology Institute, ARTI–21CR/610–40030–01. [www.osti.gov/energycitations/servlets/purl/810964-SRS2Dd/native/810964.pdf]

Weber AM, Boudreau Y [2000]. Closeout letter of August 9 to Fulton County Department of Health and Wellness, Atlanta, GA. Cincinnati, OH: U.S. Department of Health and Human Services, Centers for Disease Control and Prevention, National Institute for Occupational Safety and Health, Division of Surveillance, Hazard Evaluations and Field Studies, HETA No. 99–0034. Unpublished.

Wells MW, Holla WA [1950]. Ventilation in the flow of measles and chickenpox through a community: progress report, Jan. 1, 1946 to June 15, 1949, airborne infection study, Westchester County Department of Health. J Am Med Assoc *142*(17):1337–1344.

Wells WF [1942]. Radiant disinfection of air. Arch Phys Ther *23*:143–148.

Wells WF [1955]. Biophysics of droplet nuclei disinfection. In: Airborne contagion and air hygiene: an ecological study of droplet infections. Cambridge, MA: Harvard University Press, pp. 62–78.

Wells WF, Wells MW [1936]. Air-borne infection sanitary control. J Am Med Assoc *107*:1805–1809.

Wells WF, Wells MW, Wilder TS [1942]. The environmental control of epidemic contagion—I: an epidemiologic study of radiant disinfection of air in day schools. Am J Hyg *35:97*–121.

Westinghouse [1982]. Sterilamp germicidal ultraviolet tubes. Bloomfield, NJ: Westinghouse Lamp Commercial Division, Booklet A–8968.

WHO (World Health Organization) [2007]. Tuberculosis: fact sheet no. 104. [www.who.int/mediacentre/factsheets/fs104/en]. Date accessed: June 2007.

Willmon TL, Hollaender A, Langmuir AD [1948]. Studies of the control of acute respiratory diseases among naval recruits—I: a review of a four-year experience with ultraviolet irradiation and dust suppressive measures, 1943 to 1947. Am J Hyg *48*:227–232.

Xu P [2001]. Ultraviolet germicidal irradiation for preventing infectious disease transmission [Thesis (Ph.D.)]. Boulder, CO: University of Colorado, Department of Civil, Environmental, and Architectural Engineering.

Xu P, Miller SL [1999]. Factors influencing effectiveness of ultraviolet germicidal irradiation for inactivating airborne bacteria: air mixing and ventilation efficiency. In: Raw G, Aizlewood C, Warren P, eds. Proceedings of Indoor Air '99': 8th International Conference on Indoor Air Quality and Climate, Edinburgh, Scotland, August 8–13, Vol. 2. pp. 393–398.

Xu P, Peccia J, Hernandez M, Miller SL [2000]. The efficacy of upper room ultraviolet germicidal irradiation in inactivating airborne microorganisms under incomplete mixing conditions. Unpublished paper presented at Engineering Solutions to Indoor Air Quality Problems, Raleigh, NC, July 17–19. [http://spot.colorado.edu/~shellym/research/Peng.html]

Xu P, Peccia J, Fabian P, Martyny JW, Fennelly KP, Hernandez M, Miller SL [2003]. Efficacy of ultraviolet germicidal irradiation of upper-room air in inactivating airborne bacterial spores and mycobacteria in full-scale studies. Atmos Environ *37*(3):405–419.

Xu P, Kujundzic E, Peccia J, Schafer MP, Moss G, Hernandez M, Miller SL [2005]. Impact of environmental factors on efficacy of upper-room air ultraviolet germicidal irradiation for inactivating airborne mycobacteria. Environ Sci Technol *39*(24):9656–9664.

Zimmer JL, Slawson RM [2002]. Potential repair of *Escherichia coli* DNA following exposure to UV radiation from both medium- and low-pressure UV sources used in drinking water treatment. Appl Environ Microbiol *68*(7):3293–3299. [www.pubmedcentral.nih.gov/picrender.fcgi?artid=126789&blobtype=pdf]

APPENDIX A — FIELD PROBLEMS NOTED IN SOME UVGI SYSTEMS

The CDC/NIOSH Health Hazard Evaluation (HHE) and Technical Assistance Program receives about 400 requests a year to conduct field investigations of potential health hazards. These requests are made by employers, workers, worker representatives, other Federal agencies, and State, foreign, and local agencies. From 1987 through 2000, the program received 14 TB-related requests for technical assistance that involved evaluations of UVGI systems [Boudreau et al. 1995; Buchta et al. 1993; Burton 1995; Burton and Martinez 1997; Decker 1993; Martinez 1995a, 1995b, 1999; Moss and Seitz 1990, 1991; Murray 1987; Seitz 1992; Seitz et al. 2000; Tubbs and Bresler 1992; Weber and Boudreau 2000]. These requests came from a variety of workplaces including hospitals (e.g., emergency rooms, waiting areas, and microbiology laboratories), neighborhood health centers, TB clinics, drug treatment centers, correctional facilities, a medical examiner's office, and a homeless shelter. A review of these requests for assistance indicates the following installation or maintenance problems were noted in at least one of the facilities during the investigations:

- Lack of preventive maintenance plans, resulting in numerous reports of dirty and/or old lamps that produced minimal UVGI levels.

- No wall switches to turn off UVGI lamps or fixtures during maintenance operations.

- UV fixture louvers bent or missing, allowing workers to look directly at the lamps.

- No warning labels on UVGI lamps or fixtures to warn of hazards from direct eye contact.

- No worker training on the potential hazards of UVGI.

- No lockout mechanisms or automatic shut-off mechanisms to rooms where the upper-room UVGI system was designed to be deactivated when people were present in the rooms.

- Polished metal and reflective paint surfaces that indirectly increased the UVGI level in the occupied space.

- No warning signs in areas with high UVGI levels.

- Improper planning and installation of fixtures.

APPENDIX B — RESEARCH TOOLS

Modeling

Mathematical models have been used to provide an estimate of the UVGI irradiance in the upper room. For example, Rudnick [2001] used a mathematical model to predict the UV fluence rate at any location in the room from a multi-louvered ceiling-mounted pendant fixture. Experimental measurements of irradiance determined with a radiometer showed fairly good agreement with the fluence rate derived from the mathematical model. Nicas and Miller [1999] used a mathematical model to estimate the average upper-room fluence rate of UVGI lamp fixtures used by Riley et al. [1976]. Beggs and Sleigh [2002] developed a model that provided an average fluence rate close to the experimentally derived value reported by Miller and Macher [2000]. Computer models that are commercially available may potentially be used to estimate the average UV fluence rate produced by several fixtures in a room or area (e.g., multiphysics models).

Another theoretical approach that has been used to estimate the UV fluence rate of an upper-room UVGI system and the potential UVGI dose received by airborne microorganisms involves computational fluid dynamics (CFD) modeling [Memarzadeh 2000; Xu 2001]. Commercially available CFD models can estimate the path of a microbial particle(s) in a room from the time it enters until it is removed from the air (e.g., vented, impaction on a surface). The CFD model must be joined to a model that estimates the UV fluence that a particle receives as it passes through a room to estimate the dose received by the particle. Also, the sensitivity (e.g., Z-value) of the microbial particle to UVGI can be incorporated into the model to provide a reasonable estimate of the system efficacy.

Mathematical and CFD models used for estimating the effectiveness of upper-room UVGI systems are dependent on the assumptions on which they are based. As these models become more refined, they will increasingly become more valuable for designing upper-room UVGI systems.

Chemical Actinometry

UV fluence rate can be estimated using chemical actinometry to measure the change UVGI produces in a chemical. The chemical (spherical) actinometry method devised by Rahn uses the following photochemical reaction to measure UV irradiance [Rahn 1997; Rahn et al. 1999]:

$$3I^- + N_2O + H_2O + h\nu \text{ (photon)} \rightarrow I_3^- + N_2 + 2\,OH^-.$$

The formation of the triiodide is proportional to the UV irradiance received, and it alters the absorbency of the solution. A solution of KI with buffer was placed into specially made 0.5 cm^3 spherically shaped quartz actinometry cells. The solution used to make this type of measurement is transparent to wavelengths greater than 330 nm and opaque to wavelengths less than 290 nm. The spherical geometry of the actinometry cells was chosen to best simulate irradiation to which airborne microorganisms may be subjected. Actinometry cells can be suspended at various locations in a room to best estimate the total fluence level.

Actinometry can be used to estimate the total UV fluence rate received by airborne microorganisms; however, there are drawbacks to its application. At present, this measurement approach requires the use of a spectrophotometer, chemical mixtures, and specifically designed measurement cells.

www.ingramcontent.com/pod-product-compliance
Lightning Source LLC
Chambersburg PA
CBHW081838170526
45167CB00007B/2837